AF601888

Fish and Aquatic Resources

The Editor

Dr. V.B. Sakhare is Head, Post Graduate Department of Zoology, Yogeshwari Mahavidyalaya, Ambajogai. He has 18 years' experience as an outstanding teacher and researcher. He has done pioneering work in the field of Reservoir Fisheries and Limnology. Dr.Sakhare has successfully organized *National Conference on Emerging Trends in Fisheries and Aquaculture (ETFA-2012), National Workshop on Techniques of Scientific Writing (TSW-2014), National Conference on Current Perspectives in Limnology* (*NCCPL-2009*) and *Regional Workshop* on *Water Quality Assessment* (*Implications in Potability, Productivity and Pollution Control*).

Dr.Sakhare has been editing an international journal *'Ecology and Fisheries'* (ISSN 0974-6323).Dr.Sakhare has authored/edited few books such as *'Fish and Fisheries of Indian Reservoirs', 'Applied Fisheries', 'Reservoir Fisheries and Limnology', 'Reservoir Fisheries and Ecology: A Literary survey', 'Methodology for Water Analysis', 'Aquatic Ecology', 'Aquatic Biology and Aquaculture', 'Inland Fisheries', ' Applied Ecology', 'Perspectives in Ecology'*, and *'Advances in Aquatic Ecology* (*Vols.* I–IX).

Dr.Sakhare is a recognized post graduate teacher and research guide of Dr.Babasaheb Ambedkar Marathwada University, Aurangabad, Solapur University, Solapur and J.J.T.University, Rajasthan. Under his guidance three students have completed Ph.D. He has published more than 50 research articles and reviews in peer reviewed journals.

Dr. B. Vasanthkumar is presently working as Associate Professor and Head of the Post Graduate Department of Zoology at Government Degree Arts and Science College, Karwar, Karnataka. He is recipient of summer research fellowship of Indian Academy of Science for year 2014.He has published more than 50 research papers in the national and international journals of repute and more than 50 popular science articles in different aspects of ecology and environment. He has also published 12 textbooks for under graduate students of Karnataka University, Dharwad. He is co-author of reference books like 'Aquatic Ecosystem and its Management' *'Applied Ecology' 'Advances in Aquatic Ecology* (*Volumes 7,8 and 9*)*'* and *'Emerging Trends in Fisheries and Aquaculture'*. Presently Dr.Vasanthkunar is working as Principal Investigator with Major Research Project funded by University Grants Commission.

Dr. C.M. Bharambe is working in the Department of Zoology, Vidnyan Mahavidyalaya, Malkapur (Maharashtra). He has done significant work in the field of Limnology and Toxicology and published more than 35 research papers. Dr. Bharambe is a recognized research guide of Sant Gadge Baba Amravati University and Algappa University, Tamil Nadu. Dr. Bharambe is also life member of Indian Science Congress and Indian Association of Aquatic Biologists.

Dr. Bharambe has chaired a number of sessions of different seminars/symposia. He has been invited to different colleges/institutes to deliver lectures on different topics in the field of limnology and toxicology.

Fish and Aquatic Resources

— *Editors* —

V.B. Sakhare
B. Vasanthkumar
C.M. Bharambe

2017
Daya Publishing House®
A Division of
Astral International Pvt. Ltd.
New Delhi – 110 002

Publisher's Note:

Every possible effort has been made to ensure that the information contained in this book is accurate at the time of going to press, and the publisher and author cannot accept responsibility for any errors or omissions, however caused. No responsibility for loss or damage occasioned to any person acting, or refraining from action, as a result of the material in this publication can be accepted by the editor, the publisher or the author. The Publisher is not associated with any product or vendor mentioned in the book. The contents of this work are intended to further general scientific research, understanding and discussion only. Readers should consult with a specialist where appropriate.

Every effort has been made to trace the owners of copyright material used in this book, if any. The author and the publisher will be grateful for any omission brought to their notice for acknowledgement in the future editions of the book.

Cataloging in Publication Data--DK
Courtesy: D.K. Agencies (P) Ltd. <docinfo@dkagencies.com>

Fish and aquatic resources / editors, V.B. Sakhare, B. Vasanthkumar, C.M. Bharambe.
pages cm
Contributed articles.
Includes bibliographical references and index.
ISBN 978-93-86071-48-4 (International Edition)

1. Fishery resources--India. 2. Aquatic resources--India. I. Sakhare, V. B. (Vishwas Balasaheb), 1974- editor. II. Vasanthkumar, B., 1968- editor. III. Bharambe, C. M., editor.

QL634.I4F57 2017 DDC 333.9560954 23

Published by : **Daya Publishing House®**
A Division of
Astral International Pvt. Ltd.
– ISO 9001:2015 Certified Company –
4736/23, Ansari Road, Darya Ganj
New Delhi-110 002
Ph. 011-43549197, 23278134
E-mail: info@astralint.com
Website: www.astralint.com

Preface

Fish constitutes almost half of the total number of vertebrates in the world. They exhibit enormous diversity of shape, size, biology and habitats. Of the 39,900 species of vertebrates in the world, 21,723 extant species of fish under 4,044 genera, 445 families and 50 Orders in the world. Of these, 8,411 are freshwater species and 11,650 are marine water species.

Due to anthropogenic stresses like habitat destruction, over-exploitation, wanton killing of juveniles and brood fishes, aquatic pollution, excessive water abstraction leaving inadequate water in river stretches, spread of dreaded diseases and uncontrolled introduction of exotics, a number of fishes are showing declining trends in their catches from the conventional fishing grounds and some have become threatened too. Since maintenance of fish biodiversity along with other biotic resources is prerequisite for well-being of even human beings, the current status of fish and aquatic environment and possible strategies for the conservation of fish and aquatic resources have been discussed.

With the rapid overall development and owing to ever-increasing demand of fish as food, the aquatic ecosystems are under constant pressure of man-induced stresses to the detriment of the aquatic flora and fauna.

The present book entitled *'Fish and Aquatic Resources'* comprises 23 chapters by well known experts and research workers in the field of fish and fisheries. With its application oriented and interdisciplinary approach, we hope that the students, teachers, researchers, scientists, policy makers and environmental lawyers will find this volume much more useful.

We owe our special thanks to H.A. Sayeswara and H.M. Ashashree of Sahyadri Science College(Autonomous), Shivamogga, Dr.M.M.Girkar of College of Fishery Science, Udgir, Dr.Meenakshi Jindal of CCS Haryana Agricultural University,

Hisar, Dr.V.Ravi and M.V.Radhakrishnan of Annamalai University, Parangipettai, Dr.Kiran Rasal of Central Institute of Freshwater Aquaculture, Bhubaneswar, Prof. C.Stella of Alagappa University, Thondi Campus, Prof. N.C.Ujjania of Veer Narmad South Gujarat University,Surat, Dr.M.Venkateshwarlu of Kuvempu University, Shivamogga, Dr.Shabir A. Dar of Sher-e-Kashmir University of Agricultural Sciences and Technology, Kashmir, Dr.Sachin Satam of Marine Biological Research Station, Ratnagiri, Dr.J.B.Solanki of Junagadh Agricultural University, Verval, Dr.J.L.Rathod of Karnatak University, Karwar and Prof.S.N.Padhi of Janti (Orissa).

The authors wish to express their deep sense of gratitude to shri Anil Mittal of Astral International Private Limited, New Delhi for publishing this volume.

We are grateful to our family members for helping us by several ways and encouraging us for publication of this volume.

Finally, we will always remain a debtor to all well-wishers for their blessings, without which this volume would not have come into existence.

V.B. Sakhare

B. Vasanthkumar

C.M. Bharambe

Contents

List of Contributors

Ashashree, H.M.
Department of Zoology, Sahyadri Science College (Autonomous), Kuvempu University, Shivamogga – 577 203

Bajaniya, V.C.
College of Fisheries Science, Junagadh Agricultural University, Veraval – 362265

Balai, V.K.
College of Fisheries, Maharana Paratap University of Agriculture and Technology, Udaipur – 313 001

Balkhi, M.H.
Faculty of Fisheries, Sher-e-Kashmir University of Agricultural Sciences and Technology, Kashmir – 190 006

Bhat, Farooz A.
Faculty of Fisheries, Sher-e-Kashmir University of Agricultural Sciences and Technology, Kashmir – 190 006

Brahmane, V.T.
College of Fisheries Science, Junagadh Agricultural University, Veraval – 362 265

Chakraborty, S.K.
Central Institute of Fisheries Education, Versova, Mumbai – 400 061

Chavan, B.R.
Marine Biological Research Station, Dr. B.S. Konkan Krishi Vidyapeeth, Zadgaon, Ratnagiri – 415 612

Chavda, V.M.
College of Fisheries Science, Junagadh Agricultural University, Veraval – 362 265

Chogale, N.D.
Marine Biological Research Station, Dr. B.S. Konkan Krishi Vidyapeeth, Zadgaon, Ratnagiri – 415 612

Dar, Shabir A.
Faculty of Fisheries, Sher-e-Kashmir University of Agricultural Sciences and Technology, Kashmir – 190 006

Girkar, M.M.
College of Fishery Science, Udgir – 413 517

Gulati, Rachna
Department of Zoology and Aquaculture, CCS Haryana Agricultural University, Hisar – 125 004

Jindal, Meenakshi
Department of Zoology and Aquaculture, CCS Haryana Agricultural University, Hisar – 125 004

Koli, J.M.
College of Fisheries, Shirgaon – 415 612, Ratnagiri

Kulkarni, G.N.
College of Fisheries, Shirgaon – 415 612, Ratnagiri

Kunjir, S. N.
College of Fishery Science, Udgir – 413 517

Metar, S.Y.
Marine Biological Research Station, Dr. B.S. Konkan Krishi Vidyapeeth, Zadgaon, Ratnagiri – 415 612

Nair, S.
4, Snehankit, Prakash Society, DNC Road, Ramnagar, Dombivili-E, Mumbai – 421 201

Padhi, S.N.
Reader Emeritus, Jayanti Kutira, Near Central School, Jatni – 752 050

Pagarkar, A.U.
Marine Biological Research Station, Zadgaon, Dr. Balasaheb Sawant Konkan Krishi Vidyapeeth, Ratnagiri – 415 612

Panda, Sasmita
Lecturer in Zoology, Jatni College, Jatni – 752 050

Panigrahi, G.K.
Division of Biosciences and Bioinformatics, Myongii University, South Korea

Parmar, H.V.
College of Fisheries Science, Junagadh Agricultural University, Veraval – 362 265

Parmar, P.V.
College of Fisheries Science, Junagadh Agricultural University, Veraval – 362 265

Patil, R.D.
Department of Zoology, A.S.S. and S.P.S.'s Arts, Commerce and Science College, Navapur – 425 418

Phadke, G.G.
Department of Fish Processing Technology, Karnataka Veterinary, Animal and Fisheries Sciences University, College of Fisheries, Mangalore – 575 002

Radhakrishnan, M.V.
Department of Zoology, Annamalai University, Annamalainagar – 608 002

Rasal, Kiran D.
Fish Genetics and Biotechnology, ICAR-Central Institute of Freshwater Aquaculture, Bhubaneswar – 751 002

Rathod, J.L.
Post Graduate Department of Marine Biology, Karnatak University P.G. and Research Centre, Kodibag, Karwar – 581 303

Ravi, V.
CAS in Marine Biology, Faculty of Marine Sciences, Annamalai University, Parangipettai – 608 502

Roopa, S.V.
Post Graduate Department of Marine Biology, Karnatak University P.G. and Research Centre, Kodibag, Karwar – 581 303

Sadawarte, R.K.
College of Fisheries, Shirgaon – 415 612, Ratnagiri

Sadawarte, V.R.
Marine Biological Research Station, Dr. B.S. Konkan Krishi Vidyapeeth, Zadgaon, Ratnagiri – 415 612

Sakhare, V.B.
Post Graduate Department of Zoology, Yogeshwari Mahavidyalaya, Ambajogai – 431 517

Satam, S.B.
Marine Biological Research Station, Dr. B.S. Konkan Krishi Vidyapeeth, Zadgaon, Ratnagiri – 415 612

Sawant, A.N.
Marine Biological Research Station, Dr. B.S. Konkan Krishi Vidyapeeth, Zadgaon, Ratnagiri – 415 612

Sawate, S.S.
Growel Feeds, Technical Marketing Executive (Maharashtra and Gujarat).

Sayeswara, H.A.
Department of Zoology, Sahyadri Science College (Autonomous), Kuvempu University, Shivamogga – 577 203

Sharma, L.L.
College of Fisheries, Maharana Paratap University of Agriculture and Technology, Udaipur-313 001

Shinde, K.M.
College of Fisheries, Shirgaon – 415 612, Ratnagiri

Shingare, P.E.
Dr. Balasaheb Sawant Konkan Krishi Vidyapeeth, Dapoli – 415 612 Dist. Ratnagiri

Singh, H.
Marine Biological Research Station, Zadgaon, Dr. Balasaheb Sawant Konkan Krishi Vidyapeeth, Ratnagiri – 415 612

Sofi, F.R.
Department of Fish Processing Technology, Karnataka Veterinary, Animal and Fisheries Sciences University, College of Fisheries, Mangalore – 575 002

Solanki, J.B.
College of Fisheries Science, Junagadh Agricultural University, Veraval – 362 265

Solanki, V.M.
College of Fisheries Science, Junagadh Agricultural University, Veraval – 362 265

Stella, C.
Department of Oceanography and Coastal Area Studies, Alagappa University, Thondi Campus – 623 409

Sundaray, Jitendra K.
Fish Genetics and Biotechnology, ICAR – Central Institute of Freshwater Aquaculture, Bhubaneswar – 751 002

Tank, K.V.
College of Fisheries Science, Junagadh Agricultural University, Veraval – 362 265

Temkar, G.S.
College of Fisheries, Shirgaon, Ratnagiri – 415 639

Thomas, Saly N.
Central Institute of Fisheries Technology, Willington Island, Cochin – 682 029

Ujjania , N.C.
Department of Aquatic Biology Veer Narmad South Gujarat University, Surat – 395 007

Vasanthkumar, B.
Post Graduate Department of Zoology, Government Arts and Science College, Karwar – 581 301

Venkateshwarlu, M.
Department of P.G. Studies and Research in Applied Zoology, Jnana Sahyadri, Kuvempu University, Shankaraghatta – 577 541, Shivamogga

Chapter 1

Marine Fish Biodiversity of India: Present Status

☆ *V. Ravi*

INTRODUCTION

Fishes of the world constitute with 27, 977 valid species of the known 54, 711 species of living vertebrates. Fish biodiversity are of immense value to mankind as they have long been a staple item in the diet of many peoples and today. They form an important element in the economy of many nations. However, the biodiversity of fishes of the oceans is increasingly coming under serious threat from many human activities including overfishing, use of destructive fishing methods, pollution and commercial aquaculture. Presently, the world fish stocks have come down with: fully exploited (52 per cent), moderately exploited (20 per cent), over- exploited (17 per cent), depleted (7 per cent), under exploited (3 per cent) and recovered (1 per cent). Research on fish stock assessment plays an important role as biodiversity management tool.

Types of Fish Diversity

Fishes are divided taxonomically into four living major Classes: Myxini, Petromyzontida, Chondrichthyes (cartilaginous fishes) and Actinopterygii (bony fishes). In India, 2,163 species of finfishes have been recorded from cold water (157), warm waters of the plain (454), brackishwater (182) and marine environment (1,370).

1. Taxonomic Diversity: Facts

☆ Total Vertebrates (World): 54, 711 species (including fishes, amphibians, reptiles, birds and mammals)

☆ Total fishes (World) : 27, 977 valid species

- ☆ Total Orders: 62
- ☆ Total Family : 515
- ☆ Total Genera: : 4,494
- ☆ Total Species: 27,977
- ☆ Freshwater species: 11,952
- ☆ Species using freshwater: 12,457
- ☆ More species in the Orders: Perciformes (10, 033), Cypriniformes (3, 268), Siluriformes (2, 867)
- ☆ Largest Families:: 9 (>400 species each)
- ☆ Monotypic (containing only one species): 64 families
- ☆ >100 species present: 67 families
- ☆ 151 families: only one genus (total of 587 species) Asptroblepidae (climbing catfish)- 54 species in one genus A*stroblepus*)
- ☆ Average no. of species/family: 54
- ☆ World's largest fish: Whale shark, *Rhincodon typus*(Family Rhincodontidae) TL 12.65 m and weight 21.5 tons
- ☆ World's smallest fish: Deep sea Leftvents, *Photocorynus spiniceps* (Family Linophrynidae) TL 6.5mm

Courtesy: Nelson, 2006

2. Geographic Diversity

a. Freshwater Fish Diversity

The world's freshwater habitats occur in six major zoogeographic regions such as the Nearctic region, the Neotropical region, the Palearctic region, the African/Ethiopian region, the Oriental region and the Australian region. Each region has a fairly distinct fish fauna. In India, the freshwater fish diversity harbors about 454 species in 14 major river systems sharing 83 per cent of the drainage. The important rivers in India are Ganga, Brahmaputra, Beas, Sutlej, Mahanadi, Godavari, Krishna, Cauvery, Narmada and Tapti. Of which, the Ganga river is reported with 382 species followed by Brahmaputra (126), Mahanadi (99), Cauvery (80), Narmada (95) and Tapti (57). However, several species are common to different rivers. Certain commercially important freshwtar fishes are the carps *Catla catla, Labeo rohita, L. calbasu, L. gonius, L. bata, L. fimbriatus, L. konitius, Cirrhinus mrigala, C. cirrhosa, C. reba, Puntius dubius,* and *P. carnaticus*. Catfishes are also important groups such as *Aorichthys (Mystus) aor, A. seenghala, Wallago attu, Pangasius pangasius, Salonia silondia, Bagarius bagarius, Rita rita* and *Eutropiichthys vacha* and the commercially important murrels and other fishes such as *Channa striatus, C. marulius, C. punctatus, Clarias batrachus, Heteropneustes fossilis, Anabas testudineus, Notopterus notopterus* and *N. chitala.*

b. Estuarine Fish Diversity

Estuarine region are considered as the transition zone between freshwater of the rivers and the saline water of the seas with a 0.5 to 30 ppt salinity. The major estuarine system in India are Hooghly- Matlah, Mahanadi, Godavary, Krishna, Cauvery, Narmada, Tapti and other estuaries of east and west coasts and they harbor more than 182 fishes including commercially important species like *Elops machnata, E. saurus, Mystus gulio, Nematalosa nasus, Gerres* sp., *Sillago sihama, Megalops* sp, *Polynemus tetradactylus, Eleutheronema tetradactyla, Mugil cephalus, Liza, parsia, L. tade, Epinephelus tauvina, Tenualosa ilisha, Chanos chanos, Etroplus suratensis, E. maculates, Lutjanus argentimacultus, Lates calcarifer, Tachysurus* sp., *Arius* sp.

c. Marine Fish Diversity

Defining zoogeographic regions in the world's oceans is complicated by depth, currents and geographic locales; different species occur depending on near shore, pelagic or deep-sea environments. Temperature zones divide the sea into tropical, temperate, boreal and polar regions (Table 1.1).

Table 1.1: World Ocean Water Regions and Number of Fishes Recorded (~)

Region	*Location*	*Fishes Recorded (N)*
The Indo- West Pacific Region	Tropical waters of the Indian Ocean, the western and central Pacific Ocean, and the seas connecting the two in the general area of Indonesia.	3000 to 4000
The Eastern Pacific Region	Southern Baja California to Ecuador	800
The Western Atlantic Region	Including Bermuda, Southern Florida, the Bahamas Bank, The Caribbean Sea, the tropical and temperate positions of South America	1200
The Eastern Atlantic Region	Small region along the west coast of Africa from Sengal to Angola and extending out to oceanic islands such as Ascension and St. Helena	540
The Arctic Region	Encompassing high altitude waters of Pacific and Atlantic Oceans	415
The Antarctic Region	Restricted to Antarctic waters and the surrounding Southern Ocean	274
Temperate regions	Japanese warm temperate and Californian warm temperate regions and Eastern and Western boreal regions	3000
Pelagic region	Ocean surface with 0 to 200 m depth	350
The Deep Sea	Waters deeper than 200 m depth	1950

The United Nations Food and Agricultural Organization (FAO, 2014) have produced reviews on the state of global fishery resources. In which, China dominates globally with 13.8 million metric tons annually, while India is with 3.4 million metric tons of marine fishery landings in 2012. The other countries and their catch details are given in Table 1.2 and the dominant marine fish captures of the world are in Table 1.3.

Table 1.2: Marine Capture Fisheries: Major Producer Countries

Ranking (2012)	*Country*	*Major Captures in 2012 (tons)*
1	China	13, 86,9604
2	Indonesia	5, 42,0247
3	United States of America	5, 10,7559
4	Peru	4,80,7923
5	Russian Federation	4, 06,8850
6	Japan	3, 61,1384
7	India	3, 40,2405
8	Chile	2, 57,2881
9	Viet Nam	2, 41,8700
10	Myanmar	2, 33,2790

Table 1.3: Marine Capture of World Oceans: Major Species and Genera

Sl.No.	*Scientific Name*	*Common Name*	*Family*	*Major Capture (tons)*
1.	*Engraulis ringens*	Anchoveta	Engraulidae	4, 69,2 855
2.	*Theragra chalcogramma*	Alaska Pollock	Gadidae	3, 27,1 426
3.	*Katsuwonus pelamis*	Skipjack tuna	Scombridae	2, 79,5 339
4.	*Sardinella* spp.	Sardines	Clupeidae	2, 34,5 038
5.	*Clupea harengus*	Atlantic herring	Clupeidae	1, 84,9 969
6.	*Scomber japonicus*	Chub mackerel	Scombridae	1, 58,1 314
7.	*Decapterus* spp.	Scads	Carangidae	1, 44,1 759
8.	*Thunnus albacares*	Yellowfin tuna	Scombridae	1, 35,2 204
9.	*Engraulis japonicus*	Japanese anchovy	Engraulidae	1, 29,6 383
10.	*Trichiurus lepturus*	Largehead hairtail	Trichiuridae	1, 23,5 373
11.	*Gadus morhua*	Atlantic cod	Gadidae	1, 11,4 382
12.	*Sardina pilchardus*	European pilchard	Clupeidae	1, 01,9 392
13.	*Mallotus villosus*	Capelin	Osmeridae	1, 00,6 533
14.	*Dosidicus gigas*	Jumbo flying squid	Ommastrephidae	95,0 630
15.	*Scomberomorus* spp.	Seerfishes	Scombridae	91,4 591
16.	*Scomber scombrus*	Atlantic mackerel	Scombridae	91,0 697
17.	*Strangomera bentincki*	Araucanian herring	Clupeidae	84,8 466
18.	*Acetes japonicus*	Akiami paste shrimp	Sergestidae	58,8 761
19.	*Brevoortia patronus*	Gulf menhaden	Clupeidae	57,8 693
20.	*Nemipterus* spp.	Threadfin breams	Nemipteridae	57,6 487
21.	*Engraulis encrasicolus*	European anchovy	Engraulidae	48,9 297
22.	*Trachurus murphyi*	Chilean jack mackerel	Carangidae	44,7 060
23.	*Sardinops caeruleus*	California pilchard	Clupeidae	36,4 386

The seawater surrounding the east and west coasts of India with salinity (>30ppt) is designated as marine water. Maritime states of India greatly support marine fishery resources as of the east coast (West Bengal, Odisha, Andhra Pradesh, Tamil Nadu, Puducherry) and the west coast (Gujarat, Maharashtra, Goa, Karnataka and Kerala) and islands (Andaman and Nicobar and Lakshadweep group of Islands). The total marine fish landings during 2012 in India is 3.94 million tons compared to 3.82 million tons during 2011 showing 3.4 per cent growth and this contribution was achieved from the four regions: northwest 11.5 lakh tones (29.2 per cent), southwest 13.9 lakh tons (35.1 per cent), southeast 10.1 lakh tons (25.5 per cent) and northeast 4.0 lakh tons (10.2 per cent) (Tables 1.4 and 1.5).

Table 1.4: Marine Fish Fandings of India during 2012 and their Components (per cent)

Pelagic Fish Resources (per cent)		*Demersal Fish Resources (per cent)*		*Crustacean Resources (per cent)*		*Molluscan Resources (per cent)*	
Oil sardine	35	Perches	32	Penaeid shrimps	50.7	Squids	46
Ribbonfish	11	Croakers	20	Non-penaeid prawns	33	Cuttlefishes	45
Carangids	10	Silverbellies	13	Crabs	10.5	Octopus	5
Indian Mackerels	8	Catfishes	8	Stomatopods	5.5	Bivalves	3
Other Sardines	7	Lizardfishes	6	Lobsters	0.3	Gastropods	1
Bombayduck	6	Flatfishes	6				
Anchovies	5	Elasmobranchs	5				
Other Clupeids	4	Pomfrets	4				
Tuunnies	4	Goatfishes	3				
Seerfishes	4	Others	3				
Others	3						
Thryssa	2						
Hilsa andother shads	1						

Threats to Marine Fish Diversity

There are many factors which affect life in the sea, but fishing has caused drastically the largest changes and is considered to be the major current agent of ecological disturbances in the ocean. FAO has stated that 'modern fishing is globally nonsustainable'.

Overfishing

Overfishing means catching marine creatures faster than they can reproduce and it is the main reason for the decline of fish population. Overfishing causes:

1. Massive depletion of many fishes
2. Loss of spawners
3. Declines in average sizes of fish and other marine creatures

Table 1.5: Commonly Landed Marine Fishes of Commercial importance in India

Sl.No.	Scientific Name	Common Name	Family
1.	*Alopias vulpinus*	Common thresher shark	Alopiidae
2.	*Arius tenuispinis*	Sea catfish	Ariidae
3.	*Carcharhinus* spp.	Requiem shark	Carcharhinidae
4.	*Chirocentrus dorab*	Wolf herring	Chirocentridae
5.	*Coilia dussumier*	Anchovy	Engraulidae
6.	*Coryphaena hippurus*	Dolphinfish	Coryphaenidae
7.	*Cynoglossus arel*	Tonguefish	Cynoglossidae
8.	*Decapterus russelli*	Indian Scad	Carangidae
9.	*Epinephelus diacanthus*	Spinycheek grouper	Serranidae
10.	*Euthynnus affinis*	Mackerel tuna	Scombridae
11.	*Harpadon nehereus*	Bombayduck	Synodontidae
12.	*Hilsa ilisha*	Hilsa shad	Clupeidae
13.	*Himantura imbricata*	Scaly stingray	Dasyatidae
14.	*Himantura jenkinsii*	Jenkins's whipray	Dasyatidae
15.	*Istiophorus platypterus*	Sailfish	Istiophoridae
16.	*Katsuwonus pelamis*	Skipjack tuna	Scombridae
17.	*Lactarius lactarius*	False trevallies	Lactariidae
18.	*Lethrinus mahsena*	Sky emperor	Lethrinidae
19.	*Lutjanus* spp.	Snappers	Lutjanidae
20.	*Makaira indica*	Black marlin	Istiophoridae
21.	*Megalaspis cordyla*	Torpedo scad	Carangidae
22.	*Mobula* spp.	Eagle ray	Myliobatidae
23.	*Muraenesox bagio*	Pike eel	Muraenesocidae
24.	*Nemipterus japonicus*	Threadfin breams	Nemipteridae
25.	*Nemipterus randalli*	Threadfin breams	Nemipteridae
26.	*Osteogeneiosus militaris*	Sea catfish	Ariidae
27.	*Otolithes cuvieri*	Lesser tigertooth croaker	Sciaenidae
28.	*Pampus argentius*	Silver pomfret	Stromateidae
29.	*Parastromateus niger*	Black pomfret	Carangidae
30.	*Polynemus heptadactylus*	Threadfins	Polynemidae
31.	*Priacanthus hamrur*	Bigeyefishes	Priacanthidae
32.	*Rastrelliger kanagurta*	Indian mackerel	Scombridae
33.	*Rhynchobatus djeddensis*	Guitarfish	Rhinidae
34.	*Sardinella albella*	White sardinella	Clupeidae
35.	*Sardinella fimbriata*	Fringescale sardinella	Clupeidae
36.	*Sardinella gibbosa*	Goldstripe sardinella	Clupeidae

Contd...

Table 1.5–*Contd...*

Sl.No.	*Scientific Name*	*Common Name*	*Family*
37.	*Sardinella longiceps*	Oilsardine	Clupeidae
38.	*Saurida tumbil*	Lizardfish	Synodontidae
39.	*Scoliodon laticaudus*	Spadenose shark	Carcharhinidae
40.	*Scomberomorus commerson*	Seerfish	Scombridae
41.	*Sphyraena barracuda*	Barracudas	Sphyraenidae
42.	*Sphyrna lewini*	Hammerhead shark	Sphyrnidae
43.	*Stolephorus* spp.	Anchovy	Engraulidae
44.	*Tachysurus dussumieri*	Sea catfish	Ariidae
45.	*Thryssa* spp.	Anchovy	Engraulidae
46.	*Thunnus albacores*	Yellowfin tuna	Scombridae
47.	*Thunnus tonggol*	Longtail tuna	Scombridae
48.	*Trichurus lepturus*	Ribbonfish/hairtail	Trichiuridae
49.	*Upeneus molluccensis*	Goatfish	Mullidae
50.	*Xiphias gladius*	Swordfish	Xiphiidae

4. Loss of genetic diversity
5. Genetic change toward less desirable characteristics like smaller size potentials
6. Disruption of natural communities
7. Disruption of human communities

Habitat Loss/Degradation

Damming, deforestation, diversion of water for irrigation, conversion of marshy land and small water bodies for other purposes degrade natural habitat.

Bycatch

Every fishery catches unintended, unwanted creatures, called bycatch that include nontarget and juvenile fishes, seabirds, marine mammals and any other creatures.

Human Population Growth

Coastal habitats disappear in proportion to human population growth and therefore certain important keystone species like coral reefs, mangroves, seaweeds and seagrasses if affected, will diminish the population of eggs and larvae of marine creatures and therby disturbing breeding and nursery grounds.

Fishing Conflicts

Fish conflicts, many of them violent, have erupted around the world. Norway, Iceland, the United States, Canada, Indonesia, Taiwan, Nicaragua, Russia, China, Japan, the Philippines, France Britain, Spain, India, Pakistan, Sri Lanka and many others have been involved in international fishing disputes.

Pollution

Pollution is probably the single most factor causing major decline in the population of many fish species. Industrial, sewage, and pesticide pollution cause detrimental environment to fish life in many water bodies.

Exotic Species

Exotic species are generally introduced for recreation, control of undesirable organisms, and to improve aquaculture or capture fishery productivity. They cause replacing gradually the endemic fishes.

Diseases

Among the disease caused by bacteria, fungi, viruses, parsites *etc.*, the most virulent is epizootic ulcerative disease syndrome.

Fishing Down the Marine Food Web

Present exploitation rates of many important fish stocks will be removed from the system within 25 years. This is of great concern because it implies the gradual removal of large, long- lived predatory fishes at the top of food webs and a switch to fishing for smaller, shorter- lived fish which are lower down the food web. Health and sustainability of fisheries can be assessed by monitoring trends in the average trophic levels. Once the trophic level of fishery begins to fall, it indicates the dependence on ever- smaller fish and the collapsing of larger predatory fish stocks.

Conservation of Marine Fish Diversity

Fishing can induce genetic changes through unintentional selective breeding and fishing activities can also cause evolution. Since evolution is a change in the relative frequency of genes. Some studies showed genetic changes were driven by intense fishing. Maintenance of fish biodiversity can be viewed as a requisite for the well being of human beings.

The Indian Fisheries Act of 1897 (modified in 1956) is a landmark in the conservation of fishes. Besides provisions to control and monitor gears, mesh size and observance of fishing or closed seasons, the Act also prohibits the use of explosives or poisons to indiscriminately kill fish in any water.

The fish genetic resources can be conserved by protecting an ecosystem which is broad- based, non- specific, cost- effective and relatively simplistic in approach. It may aim in general or at specific species like endangered or threatened species. It can be affected through in situ and *ex situ* methods including gene bank.

REFERENCES

CMFRI, 2014. Central Marine Fisheries Research Institute, India, Annual Report, 2012- 13, CMFRI, Kochi, India.

FAO. 2014. The State of World Fisheries and Aquaculture 2014. Rome. 223 pp.

ICAR, 2006. Fisheries and Aquaculture. ICAR, New Delhi. 755p.

Julian Roberts, 2007. Marine Environment Protection and Biodiversity Conservation. Springer, 288p.

Levin, S.A. 2001. Encyclopedia of Biodiversity. Vol 2. Academic Press. 826p.

Nelson, J.S. 2006. Fishes of the World. IV Edition. John and Wiley Sons, Inc, 601pp.

Ormond,RFG, J.D. Gage and M.V. Angel. 1997. Marine Biodiversity. Cambridge University Press. 449p.

Pauly, D. and M.L. Palomares, 2005. Fishing down marine food web: it's far more pervasive than we thought. *Bulletin of Marine Science,* 76(2): 197- 211.

Chapter 2

Diversity and Conservation of Molluscan Fauna with Special Reference to Gulf of Mannar and Palk Bay Area

☆ *C. Stella*

ABSTRACT

Recently, molluscs have been attracting the attention of scientists and planners in view of their importance. Despite their ecological and economic importance's, many species of mollusks are facing the threat of extinction. In India, there is a lack of basic information about the distribution and abundance of molluscan species especially bivalves and gastropods along the coastal habitats. Many of the coastal habitats are under the threats from development activities and the uncontrolled exploitation of natural resource. According to our available literature so far there is no authentic data for molluscan diversity and abundance among the coastal habitats of India and also no systematic and geographic studies on the molluscan fauna. So there is need to survey of molluscan species (Gastropods and Bivalves) and to study their systematic order and geographic distribution. This data will also helpful in the selection and planning of areas for conservation and also help to contribute the evaluation of biodiversity.

Keywords: *Molluscan fauna, Diversity distribution and abundance, Gastropods and Bivalves.*

INTRODUCTION

The Molluscs constitute a significant part of world fauna today. They are of major interest to man and many among them have economic importance. Man is interested in them for many reasons. About 400 million years ago, some of these molluscs, first bivalves, started inhabiting the world's freshwater streams and lakes. It took at least another 300 million years for certain gastropods to evolve to where they were capable of populating all land and freshwater habitats. Today, molluscs live in almost all parts of the world. From the deepest ocean trenches to high up on mountains, molluscs have found their niche. A conservative estimate indicated 31,643 marine, 1865 freshwater and 24,503 land mollusks making a total of about 65,000 species (Winckworth, 1932). As per other estimates the number of living species of molluscs ranged from a very modest 50,000 to 60,000. According to Subba Rao (1991), the number of species of molluscs recorded from various parts of the world vary from 80,000 to 1,00,000. It is quite likely that up to half a million species will eventually be formally discovered, since many environments and the deeper parts of the sea-floor are very poorly known even today.

Frederic Henry Gravely (1927) initiated the investigation on the littoral fauna of Krusadai Island in the Gulf of Mannar. This investigation led to the revival of the Bulletin of the Madras Government Museum for publishing the results of researches. Gravely's work (1927-1942) on molluscs helped in completing the gallery. In 1940, Satayamurthi displayed the collections in the galleries and published the outcome of his research as Guide Books. His noteworthy publications on molluscs of Krusadaiisland in the Gulf of Mannar are "Amphineura and Gastropoda"Vol.1 (Satyamoorthy,1952,56). Beginning of 20th century is the most productive and significant period in the history of Indian Malacology with the Zoological Survey of India, Central Marine Fisheries Institute and several maritime universities contributing immensely to the knowledge of the molluscan fauna. In India till today, 5070 species of molluscs have been recorded of which 3370 species are from marine realm. Gulf of Mannar and Lakshadweep are reported to have 428 and 424 species respectively. Eight species of oysters, two species of mussels,17 species of giant clams, one species of window- pane oyster and other gastropods are reported in the Gulf of Mannar (Venkataraman, 2005)

Of the seven classes in the phylum Mollusca, class Pelecypoda comprises a group of highly specialized laterally compressed animals. There are over 7000 species encompassing oysters, mussels, scallops and clams that inhabit a variety of habitats from shallow water to the deepest of the ocean floor. From India a total of 3271 numbers of mollusks are known to occur belonging to 320 families and 591 genera of which 1900are gastropods,1100 bivalves, 210 cephalopods 41 polyplacophors and 20 scaphopods. Clams are very popular among bivalves and are distributed throughout the world and have economical importance and support both commercial fishery and mariculture. Satyamurthi (1956) carried out a detailed study on the materials collected from Krusadai Island and its vicinity,comprising both spirit-preserved and dry specimens contained inthe molluscan collections of the Madras Museum. These animals included Scaphapoda, Pelecypoda and Cephalopoda. The collections were further enriched by the repeated collections made in this area by the late Mr. M. D. Crichton. This volume is a equal to the previous one on the mollusca of Krusadai Island, published as a Bulletin of the Madras Government Museum Natural History Section (1952). Rao andSundaram (1972) studied the ecology of intertidal molluscs of Gulf of Mannar and Palk Bay. Recent works by Subba Rao and Day (2000) and

Subba Rao (2003) deal with distribution of most of the molluscan fauna occurring in Indian coast including Tamil Nadu.

CMFRI Annual Report (2003-2004) mentioned about the preparation of a check list of marine molluscs of India based on the collection of cephalopods, bivalves and gastropods from the landing centers as well as from the coastal areas, inter-tidal areas and estuaries. It included 521 species of gastropods under 238 families and 298 genera; 248 species of bivalves under 85families and 115 genera; 13 species of polyplacophores under 8families; 11 species of scaphopoda under 7 families; 9 species of aplacophores under 6 families; and 201 species of cephalopods under 26 families. A total of 84 different species of gastropods belonging to 26 families and 32 genera were tentatively identified from the samples collected from the trawllanding centers of Mandapam, Kollam and Cochin. Thirty six species of bivalves belonging to 12 different families were also collected and identified. The major investigations on the bivalveswere carried out by Preston (1916), Crichton (1941) and Gravely (1941). Bivalves and gastropods of the Indian seas were reported by Ganapati and Sarma (1972). Mahadevan and Nayar (1973) studied the pearl oyster resources of India. The CMFRI scientists conducted survey and assessment on the bivalve and gastropod resources in Gulf of Mannar area in the year 2004 and reported 77 species of bivalves belonging to 2subfamilies, 5 orders and 21 families from Tuticorin coast ofGulf of Mannar.

Description of the Study Area

Gulf of Mannar in the south east coast of India extends from Rameshswaram Island in the north to Kanyakumari in the south. It has a chain of 21 island stretching from Mandapam to Tuticorin to a distance of 140 km along the coast. Each one of the island is located anywhere between 2 and 10km from the mainland. The Gulf of Mannar Biosphere reserve has an area about 10,500km^2 running along the main land coast for about 170 nautical miles including the 21 islands in the Gulf (Figure 2.1b). The total area of the island is about 623 hectares.The Gulf of Mannar considered as 'Biologists' paradise"for ithas 3600 species of flora and fauna. Gulf of Mannar is endowed with a rich variety of marine organisms includes rare cowries, cones, volutes, murices, whelks, strombids, chanks, tonnids, prawns, lobsters, pearl oysters, sea horses, sea cucumber *etc.* Coral reefs, rocky shores, sandy beeches, mud flats, estuaries, mangrove forests, seaweed structures and seagrass beds. These ecosystems support a wide variety of fauna and flora including the biosphere reserve and particularly the marine national park of the Gulf of Mannar also Gains more importance because of the alarmingly declining population of the endangered Dugong. During the Quaternary period the Palk Strait must have been originated introducing a close connection to the Southern Gulf of Mannar and to the northern Bay of Bengal. The Palk strait is being influenced mainly by the northeast monsoon. It has the length of 137km and width of approximately 6.4 – 13.7 km. The strait has strong potential of living and non – living resources (Figure 2.1a).

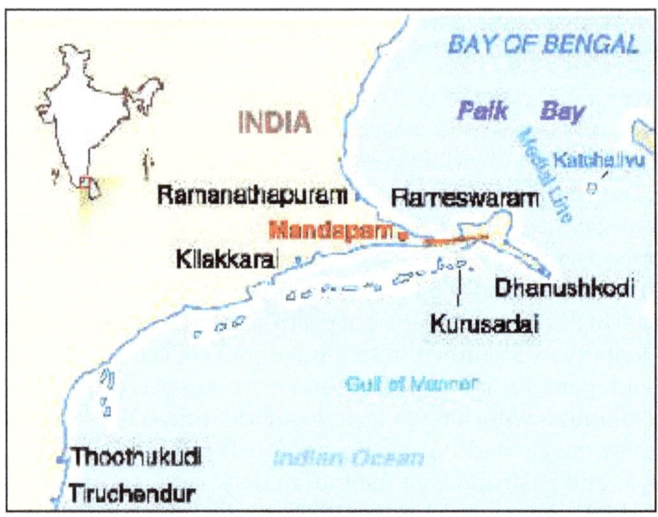

Figure 2.1a: Study Area: Palk Bay and Gulf of Mannar.

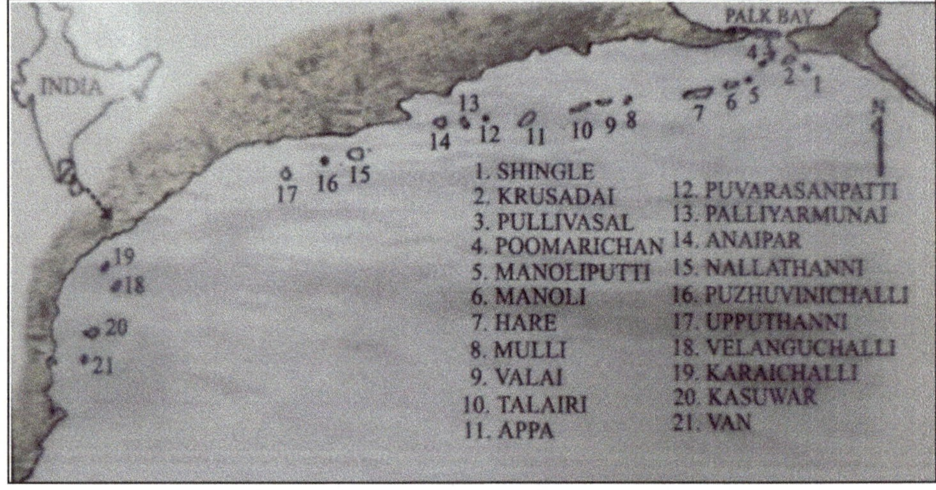

Figure 2.1b:Gulf of Mannar Biosphere Reserve Area (Mandapam and Kilakarai group of Islands).

Methodology

The Molluscan survey was performed at Mandapam and Keelakkarai group of islands. In all the fourteen islands, 7-8 transect and 4-8 quadrates of each transect were made in the Intertidal area and underwater survey to access the qualitative and quantitative analysis of scheduled molluscs and their associated fauna. The selected area is covered by dense seagrasses and seaweeds beds. The quadrate survey was made randomly in the above mentioned areas and from Shore up to 150 meters and depth range of 0.5 to 1.5 meters using snorkeling and the Under water survey, 10 to 15 meters depth using SCUBA for under water observation. The quadrates were divided into 16 grids, each 25 x 25cm followed by English *et al.* (1997).

Status and Distribution of Molluscan Fauna from Gulf of Mannar Area

During the survey (2007-2009) in the Mandapam and Kilakarai group of island, the distribution pattern of molluscs showed that the maximum number of gastropods and bivalves species were recorded in Hare island (57) and minimum number were recorded in Manoli island(41) in Mandapam group of islands and in Keelakkarai group of island,the maximum number of molluscan species were recorded in Thalaiyari island (64) and in minimum number it was recorded in Poovarasanpatty (36). The maximum number of density were recorded in Kurusadai (12.31 no/m^2) and Poomarichan island(12.61no/m^2) in Mandapam group and the minimum number it was recorded in Valimunai island(7.56 no./m^2) of Keelakkarai group.

In Mandapam group, the Kurusadai and Hare Islands showed higher density of molluscan Individuals when compared to other islands.In Keelakkarai group of islands, the Appa and Thalaiyari Island showed higher density when compared to other Islands. In both, Mandapam and Keelakkarai group, the Mandapam group showed higher density when compared to Keelakkarai group of islands. In this survey, the maximum number of species diversity was recorded in Hare Island of Mandapam group and Thalaiyari island of Keelakkarai group and in minimum number of species diversity was recorded in Manoli and Vali Munai when compared to other islands. In all the 14 islands, the gastropods diversity was higher when compared to bivalve diversity. From the observation of the survey work, so far a 132 species of gastropod from 25 families and 3 orders and 90 species of bivalves from 19 families were recorded and identified. In bivalves, out of 90 species, 30 species were newly recorded.

Status and Distribution of Molluscan Fauna in the Coastal Region from Mandapam-Valinokkam

During the survey in the coastal regions from Mandapam to Vallinokam, revealed that the distribution pattern of Molluscs from Kundukal point to Valinokkam, the scheduled species of "*Placenta placenta*" were recorded in all the 11 areas. The maximum number of scheduled species of *Placenta placenta* were recorded in Pampan (3.0 per cent), Puthumadam (3.2 per cent) and Sethukarai (3.0 per cent) and the minimum numbers of the scheduled species were recorded in Muthupettai (1.3) and Valinokkam (1.4). The percentage composition of individuals of scheduled species were recorded in all three areas.From this survey, in all the 11 areas the percentage composition of gastropod diversity was higher when it compared to bivalve diversity.

Status of Molluscan Diversity in Palk Bay

Palk Bay, named after Sir Robert Palk (1717-1798) the then Governor of Madras Presidency (1755-1763), is situated in the southeast coast of India encompassing the sea between Point Calimere (Kodikkarai) near Vedaranyam in the north and the northern shores of Mandapam to Dhanushkodi in the south. It is situated between Latitude 9° 55′ - 10° 45′ N and Longitude 78° 58′ - 79° 55′ E. The Palk Bay itself is about 110 km long and is surrounded on the northern and western sides by the coastline of the State of Tamilndau in the mainland of India. Palk Bay and Gulf of Mannar to its south are connected by a narrow passage called Pamban Strait which is about 1.2 km wide and 3 to 5 m deep that separates the Island of Rameswaram from the mainland. The Palk Bay waters merge with those of the Bay of Bengal in the northeast and the Gulf of Mannar waters in the south. The Palk Strait is just 35 km of water that is narrower than the English channel and separates the northern coast of Sri Lanka from the southeast coast of India. Therefore the international boundary line is close to the shores of both the countries. The boundary is only 6.9 km away from Dhanushkodi,11.5 km away from Rameswaram, 15.9 km away from Point Calimere, 23 km away from Vedaranyam and 24.5 km away from Thondi. Palk strait lies northeast of Palk Bay between the State of Tamil Nadu in India and the island nation of Sri Lanka and the width of Palk Bay ranges from 64 to 137 km (Cathcart, 2003). Palk Bay at its southern end is studded with a chain of submerged islands or shoals which appear to connect Dhanushkodi on Rameswaram island in Tamil Nadu and Thalaimannar on the Mannar island of Sri Lanka. This apparent bridge is also known as Ramasethu by the pious religious Hindus and has gained significance in recent days because of the Sethusamudram Ship Channel Project and the wide publicity created by the news media. This chain of shoals is known as Adam's bridge the name of which comes from the story that Sri Lanka was the site of the biblical earthly paradise and that it was created when Adam was expelled (Wikipedia, 2006). This bridge is approximately a 30 km long shallow ridge, with 9 km of islands and shallows and 21km of open water, and is of Holocene conglomerate and sandstone mantled with islands and shoals of shifting sand all of which rest upon Miocene limestone (Cathcart, 2004).The average water temperature in the Palk Bay varies from 24.6°C to 29.1°Cwith the lowest and the highest occurring in January and April respectively. The Palk Bay remains practically calm during most of the months. Turbulent conditions prevail during northeast monsoon period and freshwater streams dilute the sea near Mandapam.The coastline of Palk Bay has coral reefs, mangroves, lagoons, and sea grass ecosystems. The fishing season starts in October and lasts till February. Peak fishing season is during December to January. The annual average fish production is around85,000 tonnes. The saline water and the muddy substratum coupled with seasonal rains and discharge from Vaigai and Cauvery rivers has created a good breeding ground for pelagic and demersal fishes. It can be considered as internal waters because it is in most parts land locked and is not suitable for navigation of big ships because of shoals, currents and coral reefs. The marine environment and geographical features of the region show wide variations. The areas are rich in biological diversity and have a long history of human settlement, use and exploitation. They contain diversified and productive ecosystems such as estuaries, salt marshes, sea grass beds and mangroves that are sensitive to human activities.

During the Quaternary period the Palk Strait must have been originated introducing a close connection to the Gulf of Mannar in the south and northern Bay of Bengal in the north. The Palk straits being influenced mainly by the northeast monsoon. It has a length of 137km and width of approximately 6.4 – 13.7 km. The strait has a strong potential for both living and non-living resources. A regular survey was conducted at Palk Bay area (Lat 9 and10 and Long 79 and 80) from

Poin calimere to Mandapam. During this survey we have selected ten stations from Point calimere to Mandapam. Kodiyakarai, Vetharanniyam, Mallipatinum, Kattumavadi, Manalmelkudi, Kottaipattinum, S.P.Pattinum, Thondi, Uppur and Attankarai.The Shells were collected from the Intertidal area and fish landing center of ten stations. All the fish landing centers provide ample diversity of bivalve shells and the shells were collected manually. In Palk Bay area, we have identified and recorded as many as 156 species of bivalves belonging to 4 orders, 19 families and 49 genera and 206 species of gastropods. Among these 67 new species bivalves and 24 species of gastropods are found to be new distributional records of the Palk Bay area. The specimens of bivalve species have been deposited in the museum of Zoological Survey of India, Chennai, for future reference by references. This compilation entitled "Diversity of Bivalves – Palk Bay" having a check list of species available here, their classification and key characters for identification of bivalve species with good illustrations and a number of colour photos will serve as a ready reckoner for students and research scholars. In addition to that we have recorded 14 unidentified species from Malleidae family, 15 species from Isogonomonidae family and 16 species from strombidae family.

Summary

Documentation of the magnitude of biodiversity is an essential task for any programme involving sustainable utilization of the resource and balanced development. Reliable information about the distribution and abundance of the organisms present in each ecosystem or habitat is quite essential for the protection of biodiversity and conservation of resources. This necessitates documentation of fauna present in the important areas besides categorizing them as endemic, threatened and endangered species. This exercise in addition to information on abundance and diversity will help in the earmarking areas for conservation.

REFERENCES

CMFRI 2004. Biodiversity of marine molluscs. *CMFRI Ann. Rep.* (2003-2004) MOL/BIOD/01. pp. 55.

CMFRI 2005. Technological up-gradation of molluscan seed production. *CMFRI Ann. Rep.* (2004-2005) MF/CUL/03. pp. 52.

Crichton,M.D.1940. Marine shells of Madras.*J.Bombay.Nat.Hist.Soc.*, 42(2): 323-341. pls 1-4.

Dharmaraj, S., K. Shanmugasundaram and C. P. Suja 2005. Observations on the exploitation of clams in Tuticorin. *Mar. Marish. Infor. Serv. T and E Ser.*, 184: 10.

Gravely, F.H. 1927. Administrative report of Madras Fisheries Department for the year ending 30th March. Report No.1 of 1929. *Madras Fish.Bull.*, 23: 1-86.

Gravely, F.H. 1941. Shells and other animal remains found on the Madras Beach (Groups other than snails etc, (Mollusca 62-70). *Bull Madras Govt. Mus. New ser.*, 5: 1-112.

Gravely, F.H. 1942. Shells and other animal remains found on the Madras Beach.1.Groups other than snails *etc.* (Mollusca: Gastropoda). *Bull. Madras Govt.Mus.(New Ser.) Nat. Hist.,* 5(1): 112.

Ganapati, P.N and A.L.N. Sarma. 1972. Bivalve gastropods of the Indian seas. *Proc. Ind. Nat. Sci. Acad.,* 38B: 240-250.

Mahadevan, S. and Nayar, K.N.1973. Pearl oyster resources of India. In: Proceedings of the Symposium on Living Resources of the Seas around India. Cochin India CMFRI. pp. 659-671.

Satyamurthi, S.T. 1952. The molluscs of Krusadai Island in the Gulf of Mannar. I. Amphineura and Gastropoda. *Bull. Madras Govt. Mus., N.S., Nat. Hist. Sec.,* 1(2) Pt6: 1-265.

Satyamurti, S.T. 1956. The mollusca of Krusadai Island II. Scaphopoda, Pelecypoda and cephalopoda. *Bull. Madras. govt. Mus. New Ser. Nat. Hist. Sec.,* 1 (No.2 Pt.7): 202.

Satyanarayana Rao, K and Sundaram,K.S. 1972. Ecology of intertidal molluscs of Gulf of Mannar and Palk Bay. *Proc. Natn. Sci. Acad.,* 38B (5 and 6): 462-474.

Sahul Hameed, P and Somasundaram,S.S.N. 1998. A survey of bivalve mollusca in Gulf of Mannar, India. *Indian. J. Fish.,* 45(2): 177-181.

Subba Rao, N.V. 1991 Mollusca. *Animal Resources of India, 125-147.*

Subba Rao, N.V. and Dey,A.2000. Catalogue of marine molluscs of Andaman and Nicobar islands. *Rec.Zool.Surv.India,Occ.Paper No.,187: 1-323.*

Subba Rao, N. V., 2003. Indian seashells (part-1) poly placophora and gastropoda. *Zoological Survey of India,* Kolkata, India, 416pp.

Venkataraman, K.2005.Coastal and Marine Biodiversity of India.Indian.*Journal of Marine Science.* 34 (1): 57-75.

Victor, A.C.C., Chellam, A. and Dharmaraj, S. 1987. Pearl oyster spat collection. *CMFRI Spl. Publn.* 39: 49-53.

Winkworth. R., 1932. British marine mollusca. *J. Conchol., 19.*

Chapter 3

Fishery Resources of Karwar and Majali Area, Karnataka, Central West Coast of India

☆ *S.V. Roopa, J.L. Rathod and B. Vasanthkumar*

INTRODUCTION

Fishing, like agriculture, is a form of primary production and its history begun when man had to be content with the food what nature could provide. Initially food was simply gathered from vegetable or animal sources. So long as it was food and could be eaten. It mattered not whether it was taken from water or gathered from dry land. As the population expanded or increased day by day a need for increased food supply also increased. Hence plants begun to be cultivated and their harvesting replaced the casual collection of grains and vegetable food, through many centuries, to more or less settled life.

Then man specialized in hunting and planned fishing developed from the collection of animals by simple methods. Gradually animals stocks were more or less herded and controlled and even fishery resources were regularly harvested. This hunting of animals and forming could provide a much greater supply of food than possible from hunting the natural wild stock. This sophistication has not yet been achieved so far as sea life is concerned.

After many investigations Scientists came to know that protein is the best food for human beings. Many years man did not know about the contents in a fish. Then after several discoveries man came to know that large quantities of protein are present in fish.

For many thousands years, living creature of the sea have provided men with one of his principle source of food. This is true today and food from the sea continues to grow in importance as efforts made to establish sufficient resources to feed the multiplying world population: production being enable to keep up with growing demands for protein.

In many countries fishing had started hundreds of years ago. Initially fishing was carried out with the help of their weapons, and many gears have been used for both fishing and hunting, such as nets, hooks and some traps. Certain elements design and construction employed in catching gears are the same in hunting as in fishing.

When the term "Fishing" is used it should be remembered that fishing doesn't only mean the taking of fish. It also relates to the capture of all aquatic animals, that is not only of fish but also of crustaceans, molluscs, coelenterates, birds, mammals *etc.*

Long before when no fishing gears were invented, men used their hands along the shores of lakes, rivers and seas to capture fish and other aquatic animal.

In India prior to independence and immediately after that up to about 1950, fishing was carried out at subsistence level with the indigenous craft employing gears such as cast nets, small seines and traps operated rather close to the shore. The production was low and as mainly used for daily, subsistence and no real fish markets existed. After many years along with traditional fishery, small mechanized vessels with bottom trawl nets were introduced. After the introduction of trawler, purse-seiners were introduced a decade later.

It is well known that 71 per cent of world is covered by water and only about 29 per cent of world is covered by land. Main product of sea is fish. It is well known that at present nearly 95 per cent of the world marine fisheries catch came from the India has a 'V' shaped coast line of about 6,090 km (CMFRI, 1978). The west coast of India has about 2,400 km in length extending from Cape Comorin in the South to the Gulf of Kutch in the North. The west coast forms the western and north western border of the Bay of Bengal. The continental shelf is estimated to be about 1,00,000 sq. miles with an average width of about 35 miles. The continental shelf of India is more prominent in the western coast than in the eastern coast. The most productive demersal fisheries are located on the continental shelves of the oceans.

In both physical and chemical aspects, the water of Bay of Bengal and Arabian sea are different. The Bay of Bengal is generally less saline than the Arabian sea because of most off the rivers of Indian sub continent open to it. And upwelling and wind phenomenon occur on the S-W-coast of India. These contribute to the Arabian sea to becoming one of the rich fishing grounds in the world. Three fourth of the Indian Sea fish production in on the west coast or against one fourth on the east coast.

Fisheries resources are dynamic and renewable. Dynamism demands constant monitoring of the resources. Maintenance of renewability is the primary concern of the research on and management of fishery resources.

As far as the Fishery resources are concerned in India, marine fish catch of India arose from 0.58 million tons in 1950 to 1.075 million tons in 1970, registering about 100 per cent increase during this period. A steady rate of income in yield was maintained in the seventies with the total production of 1.42 million tons of marine fish in 1975. The yields in 1976 and 1977 were, however, 1.35 and 1.26 million tons respectively. This shows that there was a continental shelf area because of high organic productivity in this area.

As far as the fishery resources are concerned the total world annual catch of marine fish steadily rose from 17.7 million tons 1948 to 61.1 m. tons in 1970, (FAO, 1980). There was a sharp fall world catch in 1972 with an annual world total catch being only 56 per cent million tons. This fall was largely due to the sudden crash of the Peruvian Anchovy fishery from 12.6 MT. in 1970 to 4.8 million tons in 1972. However in 1976 the world catches touched an all time high of 63.1 million tons by a slow rate of increased other fishers.

During 1985 world total annual catch was 86.3 million tons reason for increasing the catch, mechanized gears and crafts are introduced in all over the world. And during last 1992 world catch was 98.11 million tons sharply increase from previous year. But so far during 1989 world record catch highest which was 100.33 million tons (FAO, 1980).

It has been estimated that in 1976, Indian Ocean has an estimated area of 75 million square km, but yielded only 0.45 kg/has against 1.8 kg/ha in pacific and 2.56 kg/ha in Atlantic. Of the total marine catch of fish a substantial amount (about 40 per cent) is consumed in fresh *i.e.*, in raw condition. The remaining is utilized by preserved and processed methods.

It has been admitted all around that a considerable portion of future addition to the world marine fish production may have to came from culture fisheries of brackish waters and estuaries, swaps, mangroves, costal lagoon areas, *etc.* decline in catch during 1976 and 1977 as compared to 1975. This is because prior to 1950, the fishing activities carried out by indigenous crafts and gears were confined to near shore waters and this sector was the major contributor to the fish production of the country.

The technological advancement and capabilities resulted in the extension of fishing operation from the near shore waters to about 40-50 m depth zone on the continental shelf, increase of fishing effort and fish production. With further introduction of mechanized boats like trawlers in 1961-62, demersal resources including shrimp, were caught which had to an increase in the marine production.

The irregular fluctuations the marine fisheries, had shown, a high rate of increase in 1970 and more or less stable trend till 1971 which dropped in 1972 and again picked up-to the earlier level with slightly rising trend from 1973 to 1977. This clearly shows the irregular fluctuation in marine fish landings. With the introduction of purse-seiners in the seventies the catches increased rapidly. At the same, growth of fishing potential by way of increase in number of active fishermen, their boats and gears have also resulted. Also the increase in catch is due to an overall increase of about 50 per cent in marine fisherman population as observed in fifteen years from 1961-62 to 1976-77.

The gear pattern greatly varies from state to state. The famous Rampani net, a large shore seine employed for shoaling mackerel and oil sardine, is used only in Karnataka and Goa. Gill nets are however, most concentrated in Tamil Nadu. and Gujarat, accounting for 40 per cent of total catch. Trawls are mostly concentrated on the west coast. Mechanization of fishing fleet is more in the north –west coast of India.

There have been significant developments in the exploitation of the marine fishery resources of India since early 1970's. These on the west coast of India, include introduction of purse seines, fishing during monsoon months, multiple days fishing by mechanized vessels, and other efficient gears. While these inputs have contributed to enhancement of the marine fish production.

It is perceived as competing with the artisanal fisheries in the inshore waters and fostering resources degradation as bottom trawling, during monsoon period, which adversely affect the spawning population of fish. The increasing fishing pressure in the inshore waters and the widening social and economic imbalance between the artisanal and mechanized sectors, has led to serious conflicts between these two groups of fishermen exploiting the resources. Frequent clashes, often aggravated and inflamed by the mechanized fishing group, have finally resulted in banning of fishing operations by the latter sector in the territorial waters from June to August.

Of the average annual marine fish production of 1.6 million tons in India, about 70 per cent is landed along the west coast. However, the fishery of the west coast is characterized by distinct seasonality. Southwest monsoon period is a lean season for fishing and allied activities, throughout the west coast Post monsoon season (September to January) is generally more productive and more than 60 per cent of total catch along the west coast is obtained during this period.

The west coast of India covering an area of 0.86 million square kilometers of Exclusive Economic Zone (EEZ), contributes to 1.29 million tons of marine fish production out of the total estimated all India marine fish production of 1.78 million tons (1985-89).

The marine fishery resources potential of the Indian EEZ has been assessed to be in the order of 2.3 to 5.5 million tons by various authors (Jones, 1973); Gorge, *et al.*, 1958; Nair and Gopinathan, 1981; James, *et al.*, 1989.

Indian EEZ covers an area of about 2.018 million sq. km region wise, the EEZ off west coast forms about 42.6 per cent, off east coast about 27.8 per cent around Andaman and Nicobar Island about 29.6 per cent of Indian EEZ. From the results obtained in the survey conducted by fishery Survey of India (FST) during the past two decades the fishery potential of Indian EEZ is assessed as 3.92 million tones of this.

The demersal stocks form about	1.93 million tonnes.
Pelagic stocks form about	1.74 million tonnes.
Oceanic resources form about	0.25 million tonnes.

Coast-wise is Considered

West coast supports	60.1 per cent of the resources
East coast supports	27.8 per cent of the resources
Lakshadweep sea	1.6 per cent of the resources
Andaman and Nicobar seas.	4.1 per cent of the resources
Oceanic waters	6.3 per cent of the resources

Depth-wise is Considered

Coast segment within 50 m. depth	58.1 per cent
Out shelf areas (50-200 m. depth)	34.9 per cent

And the rest deep sea and Ocean regions.

Resources- wise (Production in tonnes)

The major demersal stocks in the offshore and deep sea areas are the

Nemipterus sps.	110.6 tonnes
Cat fish	63.4 tonnes
Priacanthus sps.	54.8 tonnes

The potential pelagic stocks in the offshore waters are the coastal

Ribbon fish	266 tonnes
Carangids	231 tonnes
Oil sardine	181 tonnes
Mackerel	77 tonnes
Pelagic sharks	63 tonnes

Among the different groups, clupeid fishes occupy an unique position in the Indian fishery. They nearly 1/3rd of the total marine fish production of the country. The present fishing fleet of the country consists of over two lakh fishing units, mainly consisting mechanized boat of about 22,900, motorized traditional craft about 15,300, non- motorized traditional craft about 1,68,000. 63.61 per cent of the fish production is landed by the mechanized boat whereas the traditional sector accounts for 32.7 per cent catch by the deep sea trawlers is estimated to about 1 per cent.

At present Indian marine fish catch increased from 1980 to 1990. During 1981 Marine fish product is 2.44 million tons sharply decrease to 2.36 in the year of 1982. From 1982 onwards it is sharply increased to 1990. 1990 marine fish production of India was 3.79 million tons. (According to FAO data) From 1990 onwards the trend is decreasing. Now India ranks 7th among the major fishing nations of the world.

As per as Karnataka fisheries is considered, it has a coast line of 270 km and a shelf area of 25,000 sq. km. The state consist of 3 districts namely Dakshina Kannada, Udupi and Uttar Kannada. During the early fifties only indigenous crafts using gears such as cast nets, small seines and traps were operated rather close to the shore. During 1960 trawlers were operated along with traditional fishing. Due to this, fish production increased. In 1977 introduction of purse –seiners increased the pelagic fishery exploitation which has immensely increased the fish catch. During that there are 120 purse-seiners introduced along the Karnataka coast. The number is steadily increasing in every year.

Mackerel and oil sardine dominate the fisheries of Karnataka. The yield of mackerel fluctuated heavily from year to year with the annual catches ranging from 5.7 to 64.0 thousands tons during 1966-1976. oil sardine were the next important species with landings ranging from 11.8 to 52.7 thousand tones during 1966-1976.

About 70 per cent of the annual catch and 65 per cent of the annual income of both mechanized and non- mechanized units, along the Karnataka coast are observed during post monsoon season.

During 1988-89 state marine fish catch was 1,47,307 metric tons, which is sharply increased from the previous year. But in 1985-86 has a peak catch 2,00,666 metric tons which was record catch of the state. As per as Uttar Kannada is considered, it has a coast line of about 144 kms, extending from Majali in North and Bhatkal in South. The coast of Uttara Kannada supports rich coastal fisheries of Indian mackerel *Rastrelliger Konagurta* and sardines form the earlier days of British regime. However, with the advent of modern technology in the science of fishery, especially the introduction of trawlers by the turn of last two decades, even the demersal fishes were commercially exploited. The Uttara Kannada coast is known for presence of mackerel in abundance and is known as the "mackerel" coast. This forms the major catch from Rampani net (in the earlier times) contributing to 1/3rd of total production.

During 1962-63 total catch was 60,261 metric tonnes which is decreased to 48, 936 Metric Tonnes in 1988-89. But highest catch recorded in the year of 1985-86.

This district has 13 fish landing centers, among them Karwar is main landing center, Majali is comparatively small fish landing centre.

The estimated marine fish landings of India during the year 2005 has, provisionally, been estimated as 2.28 milliontonns registering about 12 per cent decrease against the estimated of previous year. Decrease in the catches of oil sardine, Indian mackerel, lesser sardines, croakers, ribbonfish, seer fish, penaeid prawns and cephalopods have contributed to this decline. There were marginal increase in the catches of Bombay duck, Carrangids and non penaeid prawns. Contributions from the west coast accounted for 67 per cent of total catch. The pelagic fin fishes constituted 55 per cent, demersal 26 per cent, crustaceans 15 per cent and molluscs 4 per cent of total landings. User friendly software to simulate and forecast different management implications of fish and shellfish resources has been designed (CMFRI,2005-06).

Earlier, during 1950' only indigenous craft were Rampani boats, Dugout Canoes and out – rigger boats were used. But now Rampani net has become extinct. The glass used are the yendi, small shore seines, boat seine, gill net, cast net and hook and lines.But for the first time machination was introduced in Karwar with the arrival of trawlers during 1961-62 followed by the purse – Seiners during 1975-76

(Shetty *et al.*, 1984). At present in Karwar there are about 211 trawlers for exploiting the demersal resources and 13 purse seines for exploiting the pelagic resources.

Important fisheries landed in Karwar are mackerel, sardine, scienids, carangids, flat fish, prawns, squilla *etc.*, now present annual fish catch in Karwar is about 12,000 metric tonnes.

Objectives

A good amount of the work has been carried out in the past three decades on the fin fish and shellfish fisheries of estuarine, inshore and backwaters of Karwar. There is no systematic data available regarding common fishes landed in Karwar and Majali areas.

Specific study on the abundance, distribution and diversity of fish along the Karwar and Majali was carried out. Hence to fulfill this lacunae a detailed study was undertaken in the present investigation on different species of fin fishes, their seasonal distribution etc, in relation with the various hydrological parameters at two different study stations.

1. To find out the variation in fish landing in two different stations for a period of two and half decade.
2. To evaluate the year wise marine fish catch available in entire Uttara Kannada District for a period of twenty years.
3. To study the month wise details of prawn landings and species composition for the period of two years.
4. To find out the qualitative and quantitative distribution of fin and shell fishes encountered in two selected fish landing stations.
5. To study the variations in hydrological abiotic factors like salinity, dissolved oxygen, pH, water temperature, air temperature which support the life in the water.

Based on these objectives a honest attempt has been made to project the clear images on the distribution, abundance of fin fishes in and around Karwar and Majali river during the period of January 206 to January 2007.The fishes were identified by referring the Days volumes and FAO identification key. Identification of pelagic, mid-water and benthic fishes were carried out simultaneously for their systemic study.

Station – I Karwar

Karwar is one of the major fishing centers situated (14° 50′ N latitude and 74° 03′ E longitude) in the district of Uttara Kannada of Karnataka State. The Kali river is an important river opening into the sea in Karwar. The Landing centre is situated approximately 2 K.m away from Karwar town on the Karwar Mangalore highway. There are a few islands in Karwar Bay and the important one among them are the Devgad island, Anjudiv island. The water depth in Devgad area varies from 7 – 10 fathoms. Depths of water at Baithkol fishing harbor is about 18-20 meters. The Karwar beach is about 3.3 miles long extending from the mouth of kali river to the port side towards Karwar city.

Station I: Karwar Fish Landing Centre.

Station I: Fish Landing Centre, Baithkol, Karwar.

Mechanised Crafts at Station I: Karwar Fish Landing Centre.

Station – II Majali

The fish landing Majali is located at 14° 52′ N latitude and 74° 04′ E longitude, this beach is located to the extreme north of the Karwar and marks the end of the boundary of the Karnataka coast, Majali is about 7-8 K.M from Karwar town. Majali place is one of small fish landing center, here both mechanized and non-mechanized boats are. More number of gill nets are being put in to operation in this region.

Biotic Study

Every fortnight collection was carried out for a period of 13 months from January 2006-January 2007 for both landing centers. The collection was made by random sampling method *i.e.* total boats engaged in fishing were calculated first. On that basis, random sampling was done.

Fish catch data was worked out by observing the number of baskets being transported. The total catch in kgs was found out by multiplying one basket fish to that of the total number of baskets transported. Each basket was weighing about 35-40 kgs. The Total catch, average catch, species composition and monthly variation in the availability fish also studied.

Abiotic Study

Hydrology

Fortnight water sampling was made to record the salinity, dissolved oxygen, water temperature. Air temperature was measured by using thermometer. The

Station II: Fish Landing Centre, Majali.

Station II: Plank Build Mechanized Boat at Majali.

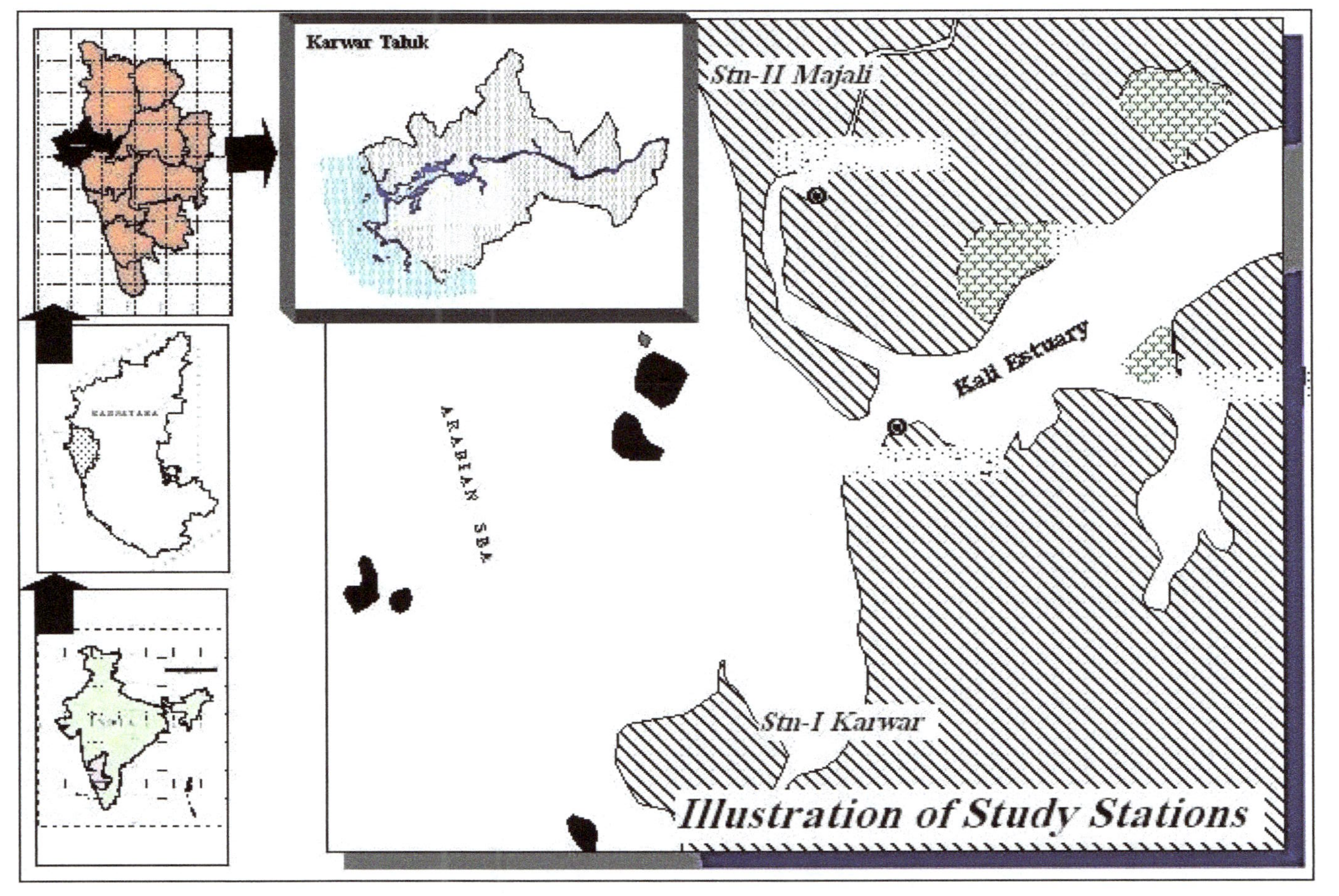

Illustration of Study Stations.

fishes were identified by referring the Days volumes and FAO identification key. Identification of pelagic, mid-water and benthic fishes were carried out simultaneously for their systemic study.

Temperature

The temperature of water and air was recorded by using centigrade thermometer nearest to 0.1°C and represented in degree centigrade.

Salinity

Salinity was determined titrimetrically by Mohr Knudsen's method (Strickland and Parson, 1975) and expressed as ppt.

pH

Portable pH meter was made use of in determination of pH in both the stations.

Dissolved Oxygen

Winkler's method as described by Strickland and Parson (1975) was followed in the estimation of the dissolved oxygen content in the water sample at study stations and recorded as mg/l of water.

There were about 251 mechanized boats like trawlers and purse – seines for the exploitation of fishes, and about 1575 non- mechanized boats are in operation at present.Hence, annual catch of Majali is very poor as compared to Karwar. Rampani was also operated during the earlier times. Today Rampani has become extinct, and smaller version of Rampani net called yendi net is usually operated now. The important fishes available are mackerel, seer fish. Prawns, flat fish, sciaenids *etc.*, There is no river opening into the sea in station II. The distance between Karwar and Majali is about 10 Kms.

Identification of Fishes fo Karawar and Majali

Vertebrates are exclusively adapted for aquatic life and have their extremities modified into fins, respiring almost invariably, solely by means of gills and possessing a two chambered heart and being cold- blooded. They are scale less partially or wholly scaled, the scales being sometimes in the form of osseous plates.

The fishes belong to the class Pisces, in this group a total 33 fin fishes were identified during study time in station (Table 3.1) and 21 types of fishes in station II (Table 3.2). The important morphological features of each fishes as well as the respective family characters as follows:

Sub class: CHONDROPTERYGII

Skeleton cartilaginous, brain distinct,skull without cranial sutures, gills pouch-like and attached by their outer edge to the skin, whilst on intervening gill-opening exists between each no gill cover. There are 5-6 pairs of gill slits.

Table 3.1: Cheklist of Finfishes Observed at Karwar (Station I)

Sl.No.	Scientific Name	Family	Popular Name	Local Name
1.	*Priacqanthus hamrur*	Priacanthide	Bulli fish	
2.	*Decapterus russelli*	Carrangidae		
3.	*Alectis indicus*	Carrangidae		
4.	*Alepes djedaba*	Carrangidae		
5.	*Megalaspis corydyala*	Carrangidae		
6.	*Caranx rotteleri*	Carrangidae		
7.	*Terapon jarbua*	Teraponidae		
8.	*Johnius voqleri*	Sciaenidae		
9.	*J. aneus*	Sciaenidae		
10.	*Pampus argenteus*	Stromateidae	Pomfrets	Silver Pomplet
11.	*Lactarius lactarius*	Lactaritidae	Buffer fish (White fish)	Sauddem
12.	*Leioqnthus dussumieri*	Leiognathidae		
13.	*Secutor insidiator*	Leiognathidae		
14.	*S. ruconius*	Leiognathidae		
15.	*Rastrellger konaqurta*	Scombridae	Mackerel	Bangada
16.	*Scomberomours linoclats*	Scombriade		
17.	*Ambasis commersioni*	Ambassidae		
18.	*Muqil cephalus*	Mugillidae		
19.	*Pempheris moluca*	Pempheridae		
20.	*Sardinella longiceps*	Clupeidae	Oil sardine	Tarli
21.	*Esualosa thoracata*	Clupeidae		
22.	*Dussumieria acuts*	Clupeidae		
23.	*Opisthopterus sps*	Clupeidae		
24.	*Sardinella fimbriata*	Clupeidae	Lesser Sardine	Pedi
25.	*Thryassa malabarica*	Engraulidae		
26.	*T. Setirostris*			
27.	*Ancohoviella* sps.			
28.	*Stolephorus devisi*			
29.	*Hemiramphus devisi*	Hemiramphide		
30.	*Arius species*	Ariidae	Cat fish	Sangate
31.	*Cynoglossus macrostomus*	Cyonglossidae	Flat fish	Lep
32.	*C. dubius*	Cyonglossidae	Flat fish	Lep
33.	*C. puncticeps*	Cyonglossidae	Flat fish	Lep

Table 3.2: Cheklist of Finfishes Observed at Majali (Station II)

Sl.No.	Scientific Name	Family Name	Common Name
1.	*Scoliodon laticaudus*	Rhinodontidae	Sharks
2.	*Himantura bleekeri*	Dasyatidae	Rays and Skates
3.	*Sardinella longiceps*	Clupidae	Oil sardine
4.	*Sardinella fimbriata*	Clupidae	Lesser sardine
5.	*Clupidae*	Clupidae	Clupidae
6.	*Silverbar*	Silverbar	Silverbar
7.	*Rastrelliger kangurta*	Silverbar	Mackerel
8.	*Scomberomorus commerson*	Scombridae	Seer fish
9.	*Thannus albacares*	Scombridae	Tuna
10.	*Lacterius lacterius*	Lactaritidae	Lactarius
11.	*Megalapsis cordila*	Carrangidae	Carrangids
12.	*Pampus argentius*	Stromatidae	Pomfrets
13.	*Anchoviella commersoni*	Engraulidae	Silver bellies
14.	*Johnious* sps.	Sciaenidae	Sciaenids
15.	*Trichiurus lepturus*	Trichiuridae	Ribbonfish
16.	*Sphyraena jello*	Soleidae	Soles
17.	*Mugil cephalus*	Mugilidae	Mullets
18.	*Metapenaeus monoceros*	Penaeidae	Brown prawn
19.	*Penaeus semisulcatus*	Penaeidae	Red prawn
20.	*Portunus pelagicus*	Nephropidae	Crabs
21.	*Shell fish*	Penaeidae	Shell fish

I. Order : PERCIFORME

1. Family : PRIACANTHIDAE

Body ovoid with ctenoid scales, mouth large and oblique, the lower jaw projecting small conical teeth in a narrow band in jaws. Eyes large, pelvic fins large originating in the advance of pectoral fins and broadly joined to body along their length a membrane. Colour usually bright red.

(i) Genus : Priacanths hamrur

Body stocky and compressed, eyes very large, mouth oblique, the lower jaw prominent. Preoperculum serrated, the spine at angle short, Gill rakers 18-24 on lower arm of first arch. The dorsal fin with 10 spines and 14 rays. Anal fin with 3 spines and 18 to 24 soft rays. Pelvic fins shorter then head. Scales small ctenoid. Colour of the body is purplish red.

Table 3.3: Catch Statistics of Karwar and Majali Fish Landing Centers (Quantity in Metric tonnes)

Sl.No.	Year	Stn. I - Karwar	Stn. II - Majali
1.	1979-80	12298	1506
2.	1980-81	9659.2	796.6
3.	1981-82	10376.2	830
4.	1982-83	10845.6	775
5.	1983-84	11893.5	1004
6.	1984-85	16552.9	953
7.	1985-86	27716	1104
8.	1986-87	15065	890
9.	1987-88	13516.5	605
10.	1988-89	9751	432
11.	1989-90	13726.6	_
12.	1990-91	80007.2	_
13.	1992-93	9827	936
14.	1993-94	9664	276
15.	1994-95	7666	300
16	1995-96	12714	544
17.	1996-97	2075.3	497.9
18.	1997-98	12449.9	382.4
19.	1998-99	16247.5	668.9
20.	1999-2000	6203.7	850.6
21.	2000-01	5764.4	986.8
22.	2001-02	6300.4	891
23.	2002-03	8895	1816.5

Priacanths hamrur

2. Family : CARRANGIDAE

Body greatly or moderately compressed, caudal peduncle rather slender. Head compressed usually keeled dorsally. Teeth is feeble. Presence of ventral scutes which is prominent or reduced in some species. Two, more or less separate dorsal fins, and soft dorsal fin with a long base.

(i) Genus : Decapterus russelli

An elongate, fusiform and moderately compressed carangid, its depth 4.2 to 4.7 times in standard length. Presence of two dorsal fins. Anal fins with two detached spines. Pecteral fins subfalcate, shorter than head length. Caudal fin deeply forked. Scutes 30-37 on leteaal line. Body bluish- green dorsally and silvery on sides and ventrally.

Decapterus russelli

(ii) Alectis indicus

Body strongly compressed and deep, its depth 1.6 to 1.9 times in standard length. Head profile almost vertical, with a marked hump above eyes. Teeth in villiform bands in jaws. Skin smooth to touch, scales minute. Lateral line strongly arched to below 9th to 12th soft dorsal rays hence straight to caudal fin base, armed wit h 5-12 weak scutes. Colour of the body pale to dark blue above, silvery white below.

(iii) Alepes djedaba

Body compressed and oblong, dorsal and ventral profiles equally and evenly convex. Eye diameter 3.5 to 4.0 times in length of head. Upper jaw(maxilla) reaching to below anterior third of eye. Gill- rankers 27-31 on lower arm of first arch. Pectoral fins falcate. Caudal fin forked. Scales on body moderate, lateral line strongly arched interiorly to below 2nd to 4th soft dorsal ray, 33 to 51 scutes. Colour of is body is bluish above, silvery below, with a distinct dusky spot on upper edge of operculum.

Other species which are identified under this family are:

(iv) Megalaspis corydyla

(v) Caranx rotteleri

Alectis indicus

Alepes djedaba

Megalaspis corydyla

Caranx rotteleri

3. Family : TERAPONIDAE

Small perch-like fishes, Body oblong or oblong-ovate, moderately compressed. Mouth usually moderate, with gape oblique. Dorsal fin nearly divided by a deep notch in some species, caudal fin rounded.

(i) Terapon jarbua

Body silvery grayish-blue, above, silvery white below, with 3 or 4 longitudinal downwardly curved stripes. Body moderately deep, teeth conic, strong, slightly recurved. Gill rackers 12 to 15 on lower arm of first arch. Dorsal fin notched. Caudal fin emarginate.

Terapon jarbua

4. Family : SCIAENIDAE

Body fairly elongate, moderately compressed, whole head and snout usually scaled. Snout rounded or bluntly pointed. eyes small or moderate, in anterior half of head. Barbel sometimes present on chin. Dorsal fin long almost completely divided by a notch, anterior part with 6-10 flexible spines and posterior part with one spine and 21-45 soft rays. Anal fin with 2 spines and 7 to 13 soft rays. Caudal fin rounded or truncate. Scales cycloid or ctenoid.

(i) *Johnius vogleri*

Greyish on back and silvery on flanks and belly. Pectoral, pelvic and anal fins yellowish. Rounded mouth but not projecting nor greatly swollen. Eyes fairly large rackets 9-12 in lower arm of first arch. Dorsal fin with 10 spines, followed by deep notch. Scales cycloid on head, ctenoid on body. Gas bladder hammer shaped with 14-15 pairs of arborescent appendages, the first entering head.

Johnius vogleri

(ii) *J. aneus*

Colour : Dark grey on back, silvery on flanks and belly. Presence of swollen snout. Gill rackers about 16 on lower arm of first arch. Caudal fin rhomboid, scales cycloid on head and breast, else where ctenoid. Gas bladder hammer-shaped with 13 or 14 pairs of arborescent appendages, the first entering head.

Johnius aneus

5. Family : STROMATEIDAE

Body deep and compressed, teeth in jaws laterally compressed, Gill membranes normal or broadly united to isthumus, the gill opening not reaching to under throat.

Single dorsal fin, anal fin usually with 2 to 6 spines and 30 to 50 soft rays. Pelvic fins absent in adults. Branchiostegeal rays 5 to 6.

(i) Pampus argenteus

Colour : Black grey, merging to silvery white towards belly, very small black dots all over body. Body very deep and compressed with firm flesh. Eye with feeble adipose lid. Single dorsal fin, falcate fin also falcate. No pelvic fins, caudal fin deeply forked the lower lobe longer.

Pampus argenteus

6. Family : LACTARITDAE

Body oblong, strongly compressed, slightly deeper than length of head, the upper and lower profiles similar. Mouth large and oblique with prominent lower jaw. Presence of two dorsal fins. Pelvic fins just below pectoral fin bases. caudal fin forked. Scales cycloid type.

(i) Lactarius lactarius

Body silvery grey, with blue iridescence above and silvery white below. Fins pale yellow. Head large, about 2.5 times in standard length, presence of canine teeth. The first dorsal fin with 7 or 8 spines, the second with 1 and 20-22 soft rays. Pectoral fins long and pointed.

Lactarius lactarius

7. Family : LELOGNATHIDAE : (Bonyfishes)

Body deep, greatly compressed and slimy, with small scales, head usually naked, the upper surface with bony ridges ending in a nuchal spine on nape. Colour silvery, often with characteristic colour markings on body and fins.

(i) Leiognathus dussueri

Body depth 2 to 2.3 times in standard length. Mouth when protracted forming a tube directed downwards, mandibular profile slightly concave or straight.

Leiognathus dussueri

(ii) Secutor insidiator

Body depth 2 to 2.3 times in standard length, mouth when protracted forming a tube directed upwards. Body silvery with blue spots on back forming vertical bands. Tip of dorsal fin black, with yellow band below.

(iii) Secutor ruconius

Body depth 1.5 to 1.8 times in standard length. Mouth small, pointing upward when protracted. Body silvery with blue spots on upper half forming about 12 vertical bands in the young specimens.

Secutor insidiator

Secutor ruconius

8. Family : SCOMBRIDAE

Mouth rather large, gill membranes free from isthmus. The dorsal fins, with finlets behind second dorsal and anal fins. Pectoral fins inserted high on body. Pelvic fin with 6 rays, placed beneath the pectoral fins. Lateral line simple or branched.

Body either uniformly covered with cycloid scales. The caudal fin rays completely cover the 'hypural" plate.

(i) ***Rastrelliger kanagurta***

Body fusiform, gill-rakers very long, visible when mouth is opened. Two dorsal fins, widely separated, first dorsal fin with 1 spine and 2 soft rays. Anal fin with 1 rudimentary spine and 2 soft rays. Scales small, ctenoid, colour- back blue-green, flanks silvery with golden tint, two rows of small dark spots on sides of dorsal- fin bases, narrow dark longitudinal bands on upper part of body and a back spot on body near lower margin of pectoral fin.

Rastrelliger kanagurta

(ii) ***Scomberomous lineolatus***

Bluish-grey on back of body the side silvery with a pattern of spots and dashes on the sides. Fine teeth on vomer and palatines. Two dorsal fins. Intestine with two folds and three distinct limbs.

Scomberomous lineolatus

9. Family : AMBASSIDAE

Body laterally compressed and somewhat translucent, operculum with a single, poorly developed spine, and double edge of the preoperrculam, so that this bone may be said to have on edge and ridge, the lower edge is nearly always dentate; dorsal fin of 2 continuous parts defined by a notch between the last and penultimate spines, pelvic fin with one strong spine and 5 soft rays, with an auxillary scale. Caudal fin forked, scale small cycloid.

(i) *Ambasis commersoni*

Body compressed, supra-orbital ridge smooth, terminating posteriorly in a single backwardly –directed spine. Suborbital entire, pre-orbital ridge entire but its edge coarsely serrated, posterior margin of pre-opercular ridge serrated with a strong spine at angle. Dorsal fin with 7 spines, followed by a deep notch the second part of fin with one spine and 9 to 11 soft rays. Body silvery with purplish reflection and bright silvery lateral band.

10. Family : MUGILIDAE

Broad, flattened head, blunt snout and cylindrical or little compressed body. Teeth absent or minute, feeble or hidden. No lateral line too short, widely separated dorsal fins. Pelvic fins sub abdominal.

(i) *Mugil cephalus*

Body olive-green above, silvery on sides, shading into white below. Head broad and much flattened on top, pre-orbital slender, filling only half the space between lip and eye, not notched, pectoral fins short, not reaching origin of first dorsal fin, with long auxiliary scale. Scales in lateral series 38 to 42. Found in estuaries, even in far away from sea, extending into almost freshwater.

Mugil cephalus

11. Family : PERPHERIDAE

Mouth usually oblique, single short dorsal fin, its origin before middle of body. Scales fairly large, easily shed.

(i) *Perpheris moluca*

Body silvery brownish on back and coppery on flanks. Body compressed deep at anterior end, narrowing sharply form anus to caudal peduncle. Mouth large, wide bank of very small granular teeth, the outer series large and conical, the anterior teeth in lower jaw partly directed forward and outwards, and there by visible from below.

Perpheris moluca

II. ORDER : CLUPEIFORMES :

Body compressed in most, swim bladder with tube-like connection to inner ear and a pneumatic duct to stomach. Branchiostegal rays usually fewer than 15. Abdomen often with keeled scutes along the ventral midline. Jaws not protrusible.

12. Family : CLUPEIDAE

Small silvery fishes, usually with fusiform sub-cylindrical bodies. Abdominal scutes usually present. Teeth small or absent. No spiny rays in fins. Dorsal and pelvic fins usually present. Pectoral fins set low on body, pelvic fins about equidistant between pectoral fin base and anal fin origin.

(i) Sardinella longiceps

Colour : Black blue-green flanks silvery, a dark spot on hind margin of gill cover. Body elongate, cylindrical, belly rounded with scutes but without prominent keel, gill rakers fine and numerous, 150 to 250 on lower arm of first arch. Dorsal fin with 13 to 15 branched rays, its origin nearer to snout –tip than to caudal fin base. Body covered with cycloid scales.

Sardinella longiceps

Other 3 species identified under family clupeidae are:

(ii) Saridinolla fimbriata

Saridinolla fimbriata

(iii) Escualosa thoracata

Escualosa thoracata

(iv) Dussumieria acuta

Dussumieria acuta

(v) *Opisthopterus* sps.

***Opisthopterus* sps.**

13. Family : ENGRAULIDAE (Anchovies)

Body usually fusiform, sub-cylindrical but sometimes, quite strongly compressed. Body sharp and rounded, with more or less numerous keeled abdominal scutes, either needle like or strongly keeled. No spiny rays in fins, dorsal fin usually short and at midpoint of body, scales cycloid. Translucent small silvery fishes.

(i) *Thryssa malabarica*

Body compressed, belly keeled with 14 to 17 pre-pelvic and 8 to 10 post-pelvic scutes, snout fairly prominent, bluntly gill-opening, not to pectoral fin base. Back brown, flanks silvery, dark venulose area on shoulder.

Thryssa malabarica

Other 3 species identified under family Engraulidae are:

(ii) *Thryssa setirostris*

Thryssa setirostris

(iii) *Anchoviella indicus*

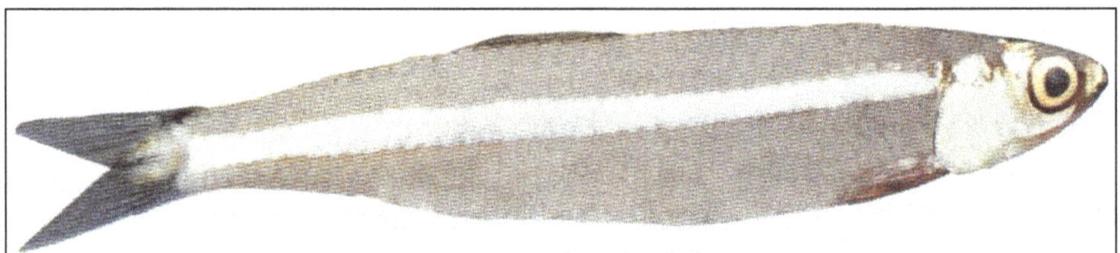

Anchoviella indicus

(iv) *Stolephorus devisi*

Stolephorus devisi

III. ORDER :ATHERINIFORMES

14. Family : HEMIRAMPHIDAE

Elongate fishes with a prolonged lower jaw and a short triangular upper jaw. Nostrils in a pit anterior to eyes. No spines in fins, dorsal and anal fin posterior in position. Pelvic fins in abdominal position, with 6 soft rays. Pectoral fin usually short. Cycloid scales, easily detached.

(i) Hemiramphus lutkei

Body rather compressed but quite robust. Upper jaw short, triangular, much broader than long, scale less. Total gill rackers. 33 to 46 on first arch, dorsal fin with 13 to 15 soft rays. Anal fin with 11 to 13 soft rays. Pectoral fin long, origin of pelvic fins twice nearest base of caudal fin than the angle of mouth. Iridescent blue, belly silvery, with lateral silvery band bordered by narrower black stripes.

IV. ORDER :GONORYNCHIFORMES

15. Family : Ariidae (Sea cat fishes)

Medium to large sized fishes, fairly elongate body without scales, caudal fin forked, some strong bony plates on head, 3 pairs of barbells-paired maxillary and mandibular barbells.Adipose dorsal fin present.

(i) Arius sps.

Body silvery steel along back, lighter on sides and below. Mouth sub-terminal its gape narrow, 3 pairs of barbels around mouth, teeth in jaw villiform. Supra-occipital process wider at its base than broad. Dorsal and pectoral fins with strong pungent spines, tip of dorsal spine prolonged into a filament.

***Arius* sps.**

V. ORDER : PLEURONECTIFOREMES (Flat Fishes)

A group of rather specialized and distinctive fishes found in most seas, which are characterized by asymmetry, primarily that of the eyes, for in the adult these lie on one side of the head, either right or left. Body highly compressed, some what convex on ocular side and flat on blind side.

Dorsal and anal fins usually long. Eyes side is always pigmented, whereas the blind side is usually white. Six or seven Branchiostegal rays, rarely eight. Body cavity small, adults almost always without swim-bladder.

Suborder : Soleodei

16. Family : CYNOGLOSSIDAE (Tongue soles)

Tongue shaped flat fishes, with eyes sinistrial (left said of head). Mouth asymmetrical, lips sometimes fringed, rostral hook present below mouth. Teeth minute and on blind side only. Pre-opercle margin not free, covered with skin and scales. Dorsal fin originates on head, both dorsal and anal fins confluent with caudal fin, pectoral fin absent. Only left pelvic fin present. No spiny rays in dorsal and pelvic fins. Scales small, ctenoid or cycloid, lateral, line variable, 0 to 3 on ocular side, 0 to 2 on blind side.

(i) *Cynoglossus macrostomus*

Body flat and elongate. Eyes nearly contiguous, with no space between them. Two nostrils on ocular side, the anterior nostril tubular and in front of lower eye, posterior nostril simple and in anterior half of inter-orbital space. Snout obtusely pointed, rostral hook short, angle of mouth reaching well beyond lower eye, nearer to tip of snout than to gill-opening. Two lateral lines on ocular side, median line with 80 to 92 scales, 14 to 16 rows of scales between them; no lateral line on blind side. Dorsal fin with 100 to 106 rays, and fin with 78 to 84 rows, caudal fin with 10 rays.

Cynoglossus macrostomus

(ii) *Cynoglossus dubius*

Eyes large, inter-orbital with pronounced. Anterior nostril of eyed side tubular, on upper lip in front of fixed eye, the posterior nostril in anterior half of inter-orbital space or in the middle of it. Snout obtusely pointed, rostral hook rather short, angle of mouth extending below vertical from posterior border of fixed eye or a little beyond, nearer to gill-opening than to tip of snout. Two lateral on ocular side. Presence of cycloid scales. Occular side uniformly brown, blind side whitish.

(iii) *Cynoglossus puncticeps*

Flat body, eyes with a narrow inter-space between them. Two nostrils on ocular side, the anterior nostril tubular and in front of lower eye, posterior nostril simple and immediately in front of lower eye, posterior nostril simple and immediately in front of inter-orbital space. Snout rounded or obtusely pointed, rostral hook short, angle of mouth not reaching beyond lower eye., slightly nearer to tip of snout than to gill-opening. Two lateral on ocular side. resence of ctenoid scales on the body. Ocular side yellow- brown with very distinct irregular dark brown blotches, often forming irregular cross-bars that disappear with age, lower side whitish, some rays of dorsal and anal fins dashed with dark brown.

The statistical analysis of selected parameters was carried out to verify their significant levels by using Sigma stat ver.3.5 a statistical package.

Cynoglossus puncticeps

Results and Discussion

The present study was carried out for a period of 13 months from January 2006 to January 207. from two different study stations *i.e.* station I (Karwar) and station II (Majali). During the present investigation an attempt was made to study the biotic and abiotic factors, from two study locales. The qualitative and quantitative study of biotic community was carried out to understand the availability of the commercially important groups of fishes. Similarly study was also carried out to know supporting abiotic factors, and the relation between biotic and abiotic factors.

Biological Parameters

During the present observation 33 verities of fishes belonging to 16 families were noticed at station I (Table 3.1). Whereas, in station II there are 21 types of fishes, representing the 15 families were recorded (Table 3.2). Blarford (1901) studied occurrence and distribution of fish and supported Day's view. Bhat (1981) studied the biology of some freshwater fishes, Dew (1981) studied the temperature influence on the abundance and the availability of marine and estuarine fishes, Rajalaxmi (1980) studied the fishery resources of Godavary estuary. Similar observation was made by Rajesh (1994) while studying the distribution, abundance and composition of fin fishes along Kali estuary. Marine fishery resources of Karwar and Majali with special emphasis on *Cynoglossus macrostomus* (Norman) was done by Raghupathi Hebbar during 1995, he too noticed similar findings.

To understand the trends in fish landings for a period of two and half decades a data was collected and pooled. A comparative study was carried out in both the study stations in the year 1979 – 80 to 2006-07. From the Table 3.3 and Figure 3.1 it can be concluded that station I is more productive than in terms of fish landing than that of station II. More numbers of active fishermen are engaged in the station I (Karwar). In station I almost 319 mechanised boats and 1304 non-mechanised boats are actively involved in fishing activity, which includes, the mechanized gears like perse seienes, trawlers, out board motor nets and gill nets.

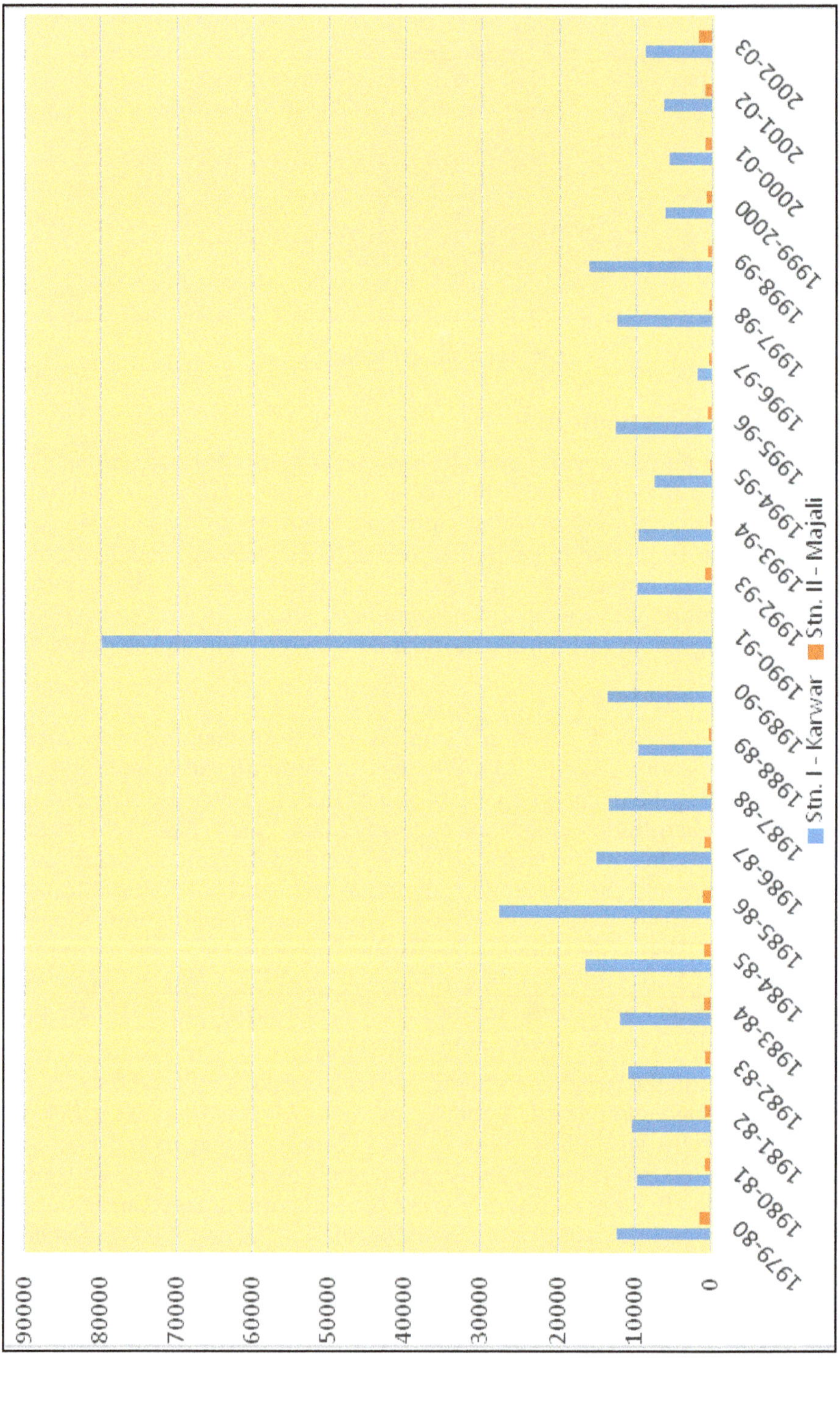

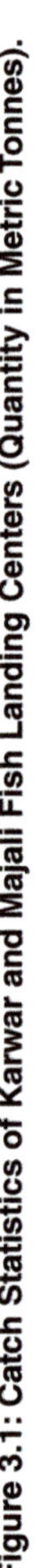

Figure 3.1: Catch Statistics of Karwar and Majali Fish Landing Centers (Quantity in Metric Tonnes).

The station II is comparative less productive because 251 mechanised boats and 1575 non-mechanised boats are engaged in fishing activity.Gill net fishing is the major gear operated in the station II.

An increasing trend of fishery was observed from 1979-80 to 1987-88 in station I, a sudden decline was recorded from 1988-89 to 1996-97, successive decreasing trend recorded in later years (Table 3.3 and Figure 3.1). Comparatively low fishing was observed in station II (Majali) than that of station I (Karwar).

The station II, showed decreasing trend in fish landing from 1979-80 to 2006-07 (Table 3.3 and Figure 3.1). Economically important fin fishes along the Kali estuary was made by Rajani Nair (2002). During her observation she too noticed same kind of fishes. Sumit Mandal (2003) carried out work on biodiversity of prosobranchs along the rocky shore of Majali coast in Uttara Kannada district.

Table 3.4: Year-wise Marine Fish Catch Quantity in M.T. of Uttara Kannada and its Value (Rs. In lakhs)

Sl.No.	*Year*	*Quantity in Metric tones*	*Value (Rs. In lakhs)*
1.	1985-86	76837	2161.56
2.	1986-87	47719	1717.99
3.	1987-88	43535	1574.33
4.	1988-89	48936	1923.70
5.	1989-90	56554	1283.91
6.	1990-91	37489	1499.41
7.	1992-93	37571	1955.09
8.	1993-94	33275	2318.50
9.	1994-95	38998	35998.34
10.	1995-96	53612	4120.16
11.	1996-97	71776.4	6165.83
12.	1997-98	47066.6	4831.54
13.	1998-99	47818.1	8217.02
14.	1999-2000	30666.4	5050.38
15.	2000-01	30095	5211.34
16.	2001-02	28037.7	4701.36
17.	2002-03	32876.65	5201.00
18.	2003-04	27574.4	4687.60
19.	2004-05	27137.1	5370.41
20.	2005-06	24517.3	6614.63
21.	2006-07	18150.71	–

Source: Fisheries Department, Govt. of Karnataka.

From the Table 3.4 and Figure 3.2, it can be stated that Uttaara Kannada district experienced decreasing trend of fish landing over period of two decades *i.e.* from the year 1985-86 to 2006-07 (Table 3.4 and Figure 3.2). The highest fish catch was recorded

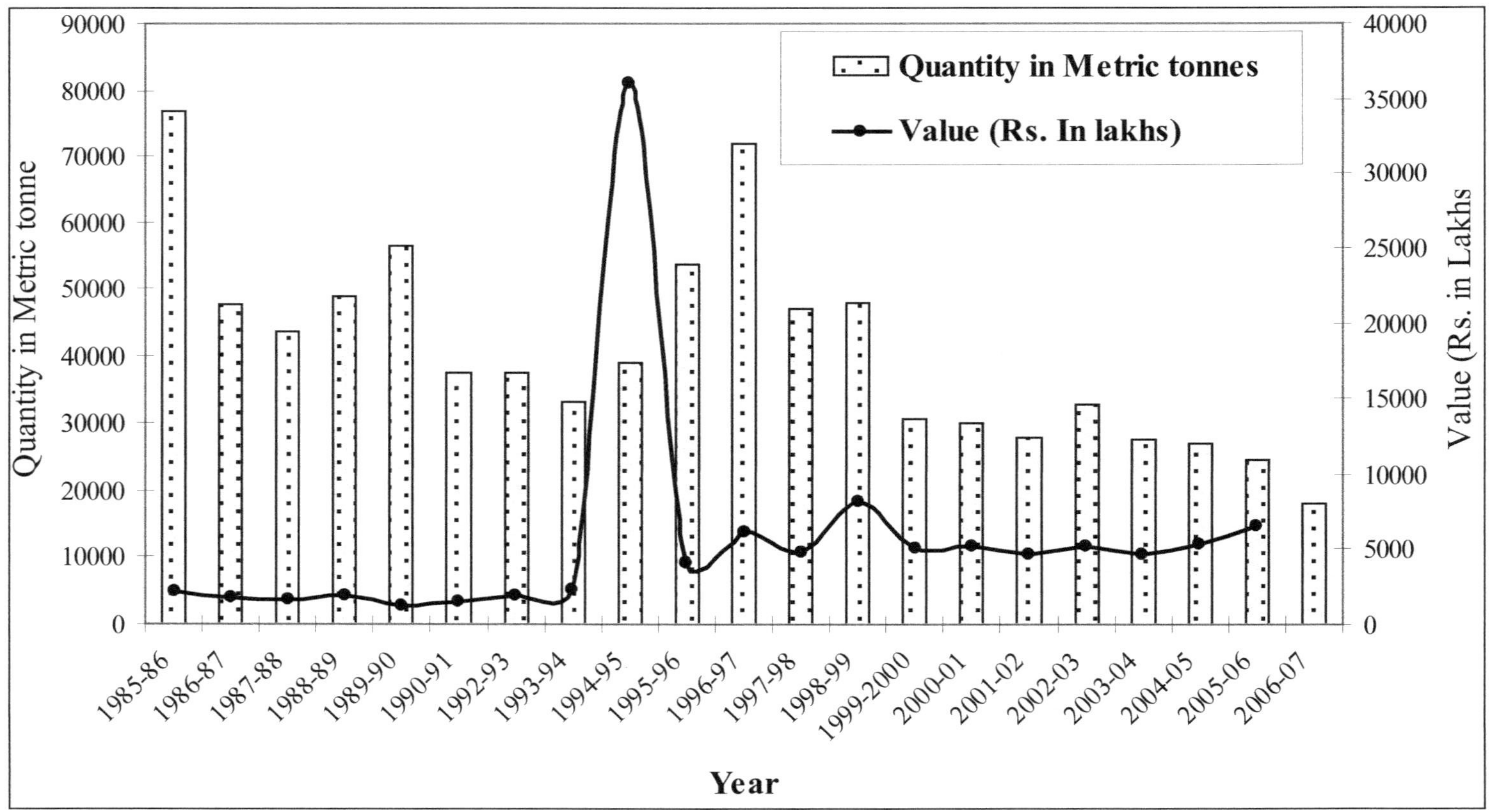

Figure 3.2: Year-wise Marine Fish Catch Quantity in M.T. of Uttara Kannada and its Value.

(71776.4 MT) value of worth of Rs. 6165.83 lakhs during 1996-97. An appraisal of the marine fisheries of Karnataka and Goa was studied by Kurup *et al.* (1977). Annigeri (1988) gave a detailed account on fish calendar of Karwar waters. Ramamurthy *et.al* (1984) made observations on the prawn fisheries of Manglore coast.

The month wise details on the species composition of prawn catch at study station I (Karwar) from the year January 2006-January 2007, consisted of the following species, *Penaeus indicus, P. semisulcatus, Metapenaeus dobsoni, Parapenaeopsis stylifera, P. merguensis* and other miscellaneous groups (Tables 3.5 and 3.6 and Figures 3.3 and 3.4).

Among prawns landed in station I (Karwar) *P. stylifera* (207324 kgs) with 36.55 per cent forming the first dominant species followed by *M. dobsoni* (164597 kgs.) with 29.02 per cent, *M.monoceros* (61246 kgs.) with 10.80 per cent, *P. merguensis* (13304 kgs) with 2.35 per cent and *M. affinis* (11183 kgs.) with lowest 1.97 per cent and other variety constituted of (109620 kgs) with 19.32 per cent (Table 3.5). Sukumaran *et al.* (1987) conducted study on the prawn fishery of Netravati Gurupur estuary Manglore stated similar kind of opinion with relevance to prawn fishery.

During the study period June – October 2006 the fish catch was not recorded because June to August it was fishing ban period, in the month of September and October 06, boats did not ventured in to the water because of lack of availability of prawns.

From the Table 3.6 it was observed that, the station I (Karwar) *M. dobsoni* (154922 kgs.) 32.08 per cent forming the first dominant species followed by *M. monoceros* (145266 kgs.) with 30.08 per cent, *P. stylifera* (120630 kgs.) with 24.98 per cent, *M. affinis* (35570 kgs.) with 7.37 per cent, *P. merguensis* (26146 kgs.) with lowest 5.41 per cent and other species variety constituted 391 kgs with.08 per cent (Table 3.6). Eco-biology and fishery of the flower tail shrimp *Metapenaeus dobsoni* (Meirs, 1878) along the Karwar water was carried out Rathod (2004) during his observation decrease trend of prawns was observed.

Table 3.7 gives the total fish catch and types of fishes available in the station I. There are 35 varities of fishes were recorded of squilla formed the major contributor to the fishery with 221 MT of catch over a period of 13 months. Oil sardine formed second important fish (160 MT) and soles formed the third group with 129 MT of catch, *Gerrus* sps. fish formed the least contributor 2 Mt.

Table 3.8 gives the total fish catch and types of fishes available in the station II. There are 35 varities of fishes were recorded of Oil sardine formed the major contributor to the fishery with 47.5 MT of catch over a period of 13 months. Shell fish formed second important fish (44.3 MT) and Scienids formed the third group with 43.5 MT of catch, Lobsters formed the least contributor 1 MT during present work.

Table 3.9 gives species composition of marine fishes available in the entire Uttara Kannada District from the year 2002-03 to 2007. Rays and skates formed the first important fish group with total of 28503.5 tonnes in five years. Followed by prawns with 8665.5 tonnes, mackerel with 8488.8 tonnes, squilla 8430.3 tonnes and least 26.7 tonnes of eels.Jowett (1990) studied the factors related to the distribution and abundance of brown and rainbow trout in New-Zealand clear water Kainz, *et*

Table 3.5: Month-wise Details of Prawn Landings (in kg) and Species Composition of Prawns (Catch and Percentage) at Station I (Karwar) Fishing Harbour in the Year 2005

	Jan	*Feb*	*Mar*	*Apr*	*May*	*Jun*	*Jul*	*Aug*	*Sep*	*Oct*	*Nov*	*Dec*	*Annual*
Prawn Catch	73088	79073	85873	159933	136355	_	_	_	_	_	22750	10202	567274
P. merguiensis	727	2374	2742	2398	4554	_	_	_	_	_	300	209	13304
Per cent	0.99	3.00	3.19	1.50	3.34	_	_	_	_	_	1.32	2.05	2.35
M. affinis	919	3272	897	3628	2467	_	_	_	_	_	_	_	11183
Per cent	1.26	4.14	1.04	2.27	1.81	_	_	_	_	_	_	_	1.97
M.monoceros	2358	4237	8820	22586	22337	_	_	_	_	_	450	458	61246
Per cent	3.23	5.36	10.27	14.12	16.38	_	_	_	_	_	1.98	4.49	10.80
M. dobsoni	23531	15654	15670	43697	51056	_	_	_	_	_	7000	7989	164597
Per cent	32.20	19.80	18.25	27.32	37.44	_	_	_	_	_	30.77	78.31	29.02
P. stylifera	42838	40946	33534	53334	20126	_	_	_	_	_	15000	1546	207324
Per cent	58.61	51.78	39.05	33.35	14.76	_	_	_	_	_	65.93	15.15	36.55
Others	2715	12590	24210	34290	35815	_	_	_	_	_	_	_	109620
Per cent	3.71	15.92	28.19	21.44	26.27	_	_	_	_	_	_	_	19.32

Table 3.6: Month-wise Details of Prawn Landings (in kg) and Species Composition of Prawns (Catch and Percentage) at Station I (Karwar) Fishing Harbour in the Year 2006

	Jan	*Feb*	*Mar*	*Apr*	*May*	*Jun*	*Jul*	*Aug*	*Sep*	*Oct*	*Nov*	*Dec*	*Annual*
Prawn Catch	110889	110073	84971	74687	20948	37319	_	7229	527	_	_	36282	482925.00
P. merguiensis	418	4962	9513	7515	1220	2012	_	149	4	_	_	353	26146.00
Per cent	0.38	4.51	11.20	10.06	5.82	5.39	_	2.06	0.76	_	_	0.97	5.41
M. affinis	522	6906	9316	5412	1868	5233	_	_	_	_	_	6313	35570.00
Per cent	0.47	6.27	10.96	7.25	8.92	14.02	_	_	_	_	_	17.40	7.37
M. monoceros	8919	33553	36110	34894	11976	15858	_	408	_	_	_	3548	145266.00
Per cent	8.04	30.48	42.50	46.72	57.17	42.49	_	5.64	_	_	_	9.78	30.08
M. dobsoni	18697	34046	30032	26866	5813	14216	_	6475	523	_	_	18254	154922.00
Per cent	16.86	30.93	35.34	35.97	27.75	38.09	_	89.57	99.24	_	_	50.31	32.08
P. stylifera	82090	30606	_	_	71	_	_	197	_	_	_	7666	120630.00
Per cent	74.03	27.81	_	_	0.34	_	_	2.73	_	_	_	21.13	24.98
Others	243	_	_	_	_	_	_	_	_	_	_	148	391.00
Per cent	0.22	_	_	_	_	_	_	_	_	_	_	0.41	0.08

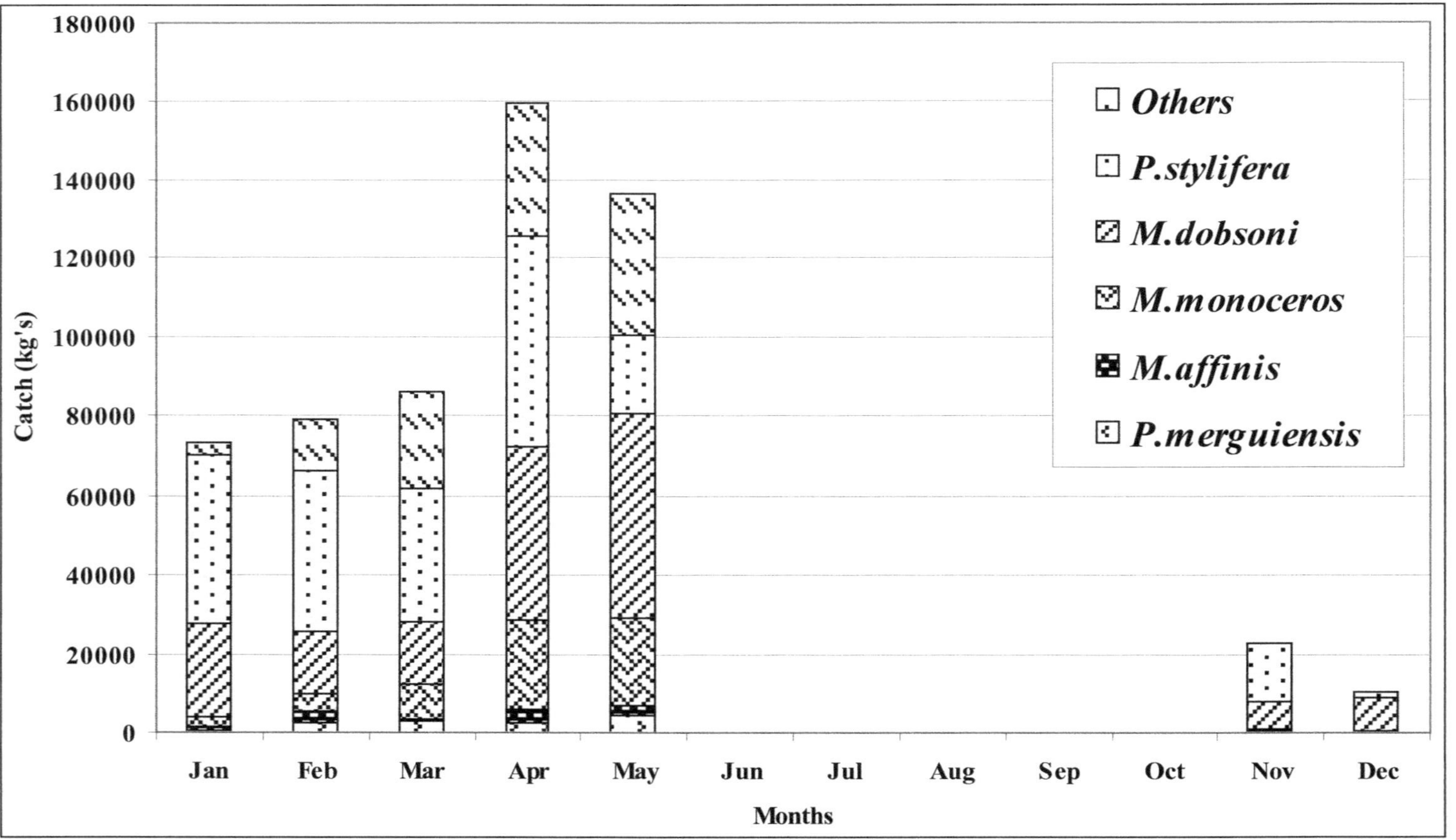

Figure 3.3: Month-wise Details of Prawn Landings and Species Composition of Prawns at Station I (Karwar) Fishing Harbour in the Year 2005.

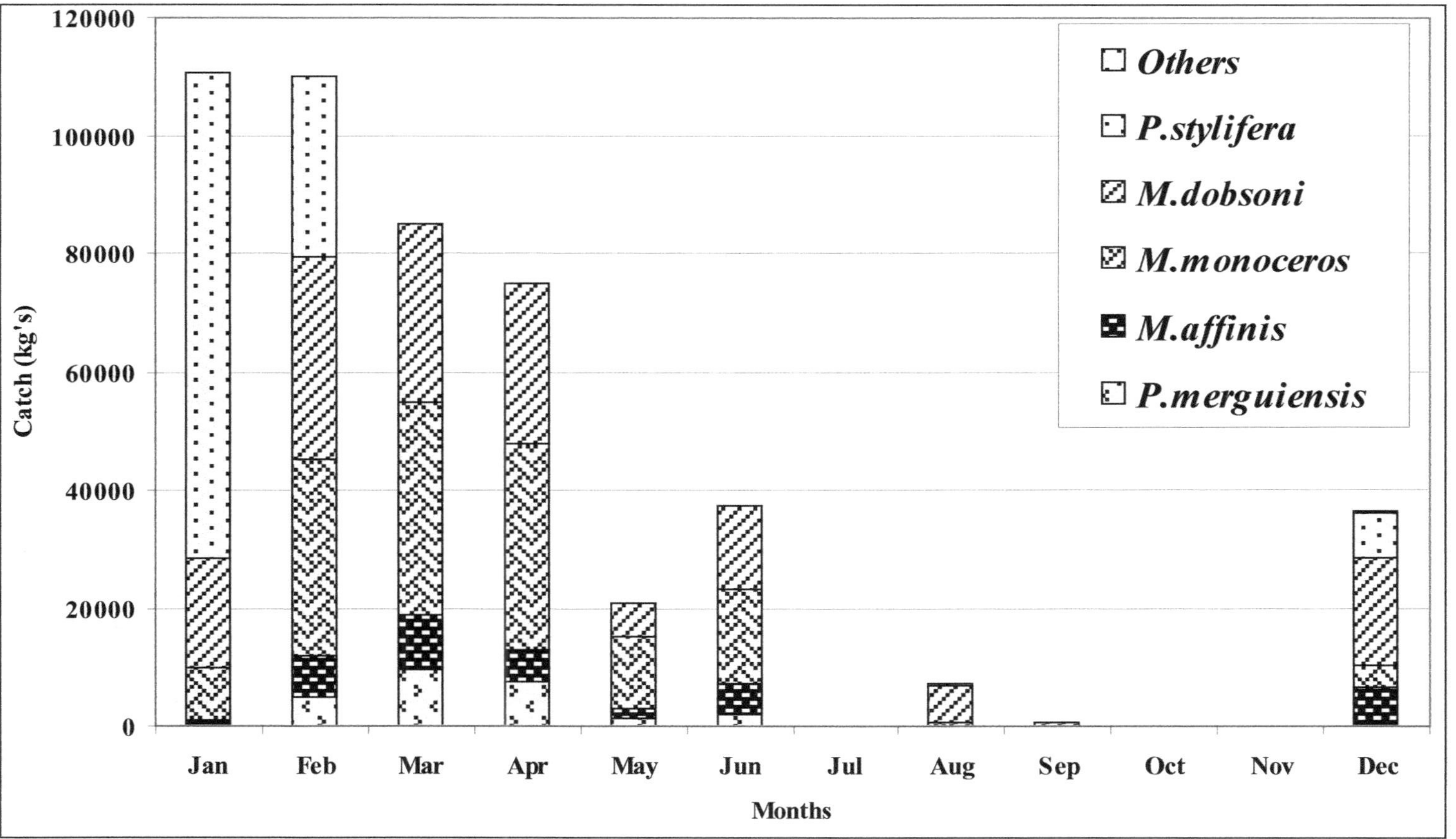

Figure 3.4: Month-wise Details of Prawn Landings and Species Composition of Prawns at Station I (Karwar) Fishing Harbour in the Year 2006.

Table 3.7: Total Fish Catch in Station-I (Karwar) from January 2006-January 07 (Quantity in Metric tonnes)

Sl.No.	Fishes	Jan-06	Feb-06	Mar-06	Apr-06	May-06	Jun-06	Jul-07	Aug-06	Sep-06	Oct-06	Nov-06	Dec-06	Jan-07
1.	Sharks				1						1	2	1	
2.	Rays and Skates	2	1								1	1	1	
3.	Oil sardine	3	7	3	12	44	5	11	5	31	16	13	9	1
4.	Lesser sardine		3		9	8	4	7	4	13	8	5	5	
5.	White sardine					2	2	8	2					
6.	Clupidae	1	4	1	15	7	6		6	17	12	8	6	1
7.	Anchovies				5	2	2		2					
8.	Silverbar		2		4	6	2		2		2	3	2	
9.	Mackerel	4	6	5	8	6	3		3		20	16	10	1
10.	Seer fish		1		7	18	6		6		3	4	2	
11.	Tuna		2		4	1	2		2		1	2	1	
12.	Lactarius		3		2		2	8	2		5	4	4	
13.	Carrangids		4		2		3	10	8	20	12	9	5	
14.	Pomfrets									16				
	a. Black pomfret		1		1		1		1		2	3	2	
	b. Silver pomfret		1		1		1		1		1	1	1	
15.	Silver bellies		3		43	18	12		12		4	3	2	
16.	Gerrus sps									2				
17.	Sciaenids	2	6		1			15			15	10	8	1.5
18.	Ribbonfish		3		13		2		2	12	5	2	4	
19.	Flatfish		1					8			2	3	1	
20.	Soles				60	28	14	6	15	4	1	1		
21.	Catfish		3		2			5		1	2	3	1	

Contd...

Table 3.7–*Contd...*

Sl.No.	*Fishes*	*Jan-06*	*Feb-06*	*Mar-06*	*Apr-06*	*May-06*	*Jun-06*	*Jul-07*	*Aug-06*	*Sep-06*	*Oct-06*	*Nov-06*	*Dec-06*	*Jan-07*
22.	Eels									4				
23.	Jawfish													
24.	Pinkperch													
25.	Bigeye													
26.	Lizardfish													
27.	Rockcod		14											
28.	Threadfin breams													
29.	Ladyfish				7		2	3	2	3				
30.	Mullets		8		2					6				
31.	Pearlspot													
32.	Lobsters					0			4					
33.	Crabs	1	2	2	16	2	4	18		7	4	5	4	0.5
34.	Squilla				190	31								
35.	MOLLUSCS													
	a. Squids		1		10		2		2			4	1	
	b. Cuttlefish										3			
	c. Octopus													
	d. Shellfish	1	8		12	3		10		5	2	5	2	1
	e. Miscellaneous	3	3.6	3	39	40	21	6	21	12	15	10	6	1
	TOTAL	**17**	**87.6**	**14**	**466**	**216**	**96**	**115**	**102**	**153**	**137**	**117**	**78**	**7**
	AVERAGE	**2.1**	**3.8**	**2.8**	**18.6**	**13.5**	**4.8**	**8.8**	**5.1**	**10.2**	**6.0**	**5.1**	**3.5**	**1.0**

Table 3.8: Total Fish Catch in Station-II (Majali) from January 2006-January07 (Quantity in Metric tonnes)

Sl.No.	*Fishes*	*Jan-06*	*Feb-06*	*Mar-06*	*Apr-06*	*May-06*	*Jun-06*	*Jul-07*	*Aug-06*	*Sep-06*	*Oct-06*	*Nov-06*	*Dec-06*	*Jan-07*
1.	Sharks	1		2	1	0.3					1			2
2.	Rays and Skates	2		1	0.2									1
3.	Oil sardine	10	1	1	5	2.5	11				1	2	2	12
4.	Lesser sardine	3		4	5	1.2	7							4
5.	White sardine													
6.	Clupidae	4	4	9	3		2	3	2	3	2	2	3	5
7.	Anchovies													
8.	Silverbar	2	1	3							2	1		1
9.	Mackerel		7	8	3	1	0.5	1			2	2	5	
10.	Seer fish	2	3	2	1	0.3					4	3	3	2
11.	Tuna	4	1	1								1	1	3
12.	Lactarius	5	2	3			0.2	2	3	1	1	1		2
13.	Carrangids	3	3	6			10		1.5	2	1	1	1	6
14.	POMFRETS													
	a. Black pomfret	1		1										1
	b. Silver pomfret	1		2										1
15.	Silver bellies	2		4	1		0.1	0.3	1	1	1	1	0.5	2
16.	*Gerrus* sps.								2	1				
17.	Sciaenids		3	7	4	3	1	4	3	2	2	2.5	4	8
18.	Ribbonfish	6	21	6										3
19.	Flatfish	1	1	1			8	3	2	2				2
20.	Soles	2					6	1	2	2	1	0.5	0.5	
21.	Catfish			4			0.2	2	1	1	1	1	2	2

Contd...

Table 3.8–*Contd...*

Sl.No.	*Fishes*	*Jan-06*	*Feb-06*	*Mar-06*	*Apr-06*	*May-06*	*Jun-06*	*Jul-07*	*Aug-06*	*Sep-06*	*Oct-06*	*Nov-06*	*Dec-06*	*Jan-07*
22.	Eels	3												
23.	Jawfish													
24.	Pinkperch													
25.	Bigeye													
26.	Lizardfish		2											
27.	Rockcod			10										
28.	Threadfin breams													
29.	Ladyfish				0.3		1							
30.	Mullets		3	2	1		0.3	0.4	1	2				
31.	Pearlspot													
32.	Lobsters	1												
33.	Crabs	2	3	4	2		0.2	0.5	3	3	2	3	2	3
34.	Squilla													
35.	MOLLUSCS													
	a. Squids	1		2										2
	b. Cuttlefish													
	c. Octopus													
	d. Shellfish	1	4	8	3	2	0.3	1	8	4	3	4	5	1
	e. Miscellaneous	3	3	6	1.5	0.7	1.2	3	3	6	2	3	6	2
	TOTAL	**60**	**62**	**97**	**31**	**11**	**49**	**21.2**	**32.5**	**30**	**26**	**28**	**35**	**65**
	AVERAGE	**2.7**	**3.9**	**4.0**	**2.2**	**1.4**	**3.1**	**1.8**	**2.5**	**2.3**	**1.7**	**1.9**	**2.7**	**3.1**

al. (1990) studied the distribution of small sized fish species of Australian running water. Kibria *et al.*(1987) conducted work on abundance and distribution of the fin fishes and prawn at five trawling stations off the Meghna River, Bangladesh.

Table 3.9: Species Composition of Marine Fish Landings in Uttara Kannada District fom 2002-2007 (Quantity in Tonnes)

Sl.No.	Species	2002-03	2003-04	2004-05	2005-06	2006-07 (Feb-07)	Total
1.	Sharks	271.8	199.2	186.1	217.3	119.7	994.1
2.	Rays and Skates	78.4	71.8	93.5	171.6	72.5	487.8
3.	Oil sardine	6590.5	7394.3	6926.2	4547.5	3045.0	28503.5
4.	White sardine	264.0	99.6	176.9	192.4	103.3	836.2
5.	Lesser sardine	683.8	539.0	736.7	1126.9	412.9	3499.3
6.	Clupidae	816.4	611.6	853.8	853.6	526.5	3661.9
7.	Silverbar	435.6	140.8	245.2	268.4	152.5	1242.5
8.	Mackerel	1759.8	1272.1	1980.8	2088.0	1388.1	8488.8
9.	Seer fish	1003.5	619.9	824.7	1059.2	520.2	4027.5
10.	Tuna	642.5	489.7	402.7	369.7	332.1	2236.7
11.	Lactarius	354.1	276.8	334.2	556.0	343.1	1864.2
12.	Ladyfish	257.6	119.2	173.6	236.1	92.7	879.2
13.	Mullets	251.9	178.4	238.9	269.4	140.5	1079.1
14.	Carrangids	1263.1	1090.4	1214.5	961.5	653.2	5182.7
15.	Pomfrets	314.1	222.6	376.9	401.9	197.6	1513.1
16.	Silver bellies	1128.8	554.7	583.4	819.7	316.3	3402.9
17.	*Gerrus* sps.	44.6	67.6	100.2	100.3	84.6	397.2
18.	Sciaenids	643.8	584.9	485.8	731.4	420.3	2866.2
19.	Ribbonfish	615.8	174.7	241.8	382.6	268.0	1682.9
20.	Flatfish	1410.8	1302.8	468.9	455.7	295.3	3933.5
21.	Anchovies	594.6	161.8	171.7	159.5	30.8	1118.4
22.	Catfish	289.3	146.7	194.6	261.3	167.8	1059.7
23.	Eels	22.9	1.0			2.8	26.7
24.	Soles	90.3	460.0	207.0	327.4	145.5	1230.2
25.	Jawfish	150.0	27.0	133.6		2060.0	2370.6
26.	Prawns	1718.1	1538.5	1977.2	2379.9	1051.8	8665.5
27.	Crabs	409.0	346.0	419.9	509.7	355.8	2040.4
28.	Shellfish	853.9	374.4	543.3	895.9	588.3	3255.8
29.	Squids	272.6	451.7	240.0	360.3	292.4	1617.0
30.	Squilla	2676.1	2419.2	1259.4	921.8	1153.8	8430.3
31.	Miscellaneous	6969.7	5638.5	5345.0	3424.2	2233.7	23611.1
	Total	**32876.7**	**27574.4**	**27137.1**	**25049.2**	**15527.7**	**128165.1**

Source: Fisheries Department, Govt. of Karnataka.

Hydrography

The study of hydrological parameters becomes significant in understanding the correlation between the biotic and abiotic factors. In this view a study was also carried out to analyse the parameters like Dissolved oxygen (D.O.) level in water, salinity, pH, water and air temperatures.

In the present observation the variations in the D.O. content of water sample in station I and II were carried out for a period of 13 months. The highest values of D.O. recorded in the month of June and July 2006 is 6.2 mg/l in station I, it is because of enough land runoff and heavy churning of water causes during monsoon and there will be more diffusion oxygen in to the water and lowest value was observed in the month February 2006 is 5.0 mg/l is due to more temperature and high salinity (Anderson *et al.*, 1984).Catching methods in fisheries, Fish catching methods of the World. pp 1-16, Third Edition 1984. Kasturirangan (1957) worked on study of the seasonal change in the dissolved oxygen of surface waters of sea in the Malabar coast. Purohit (1988) in his studies in Cochin waters stated that during pre-monsoon the dissolved oxygen values were comparatively low.

Table 3.10: Dissolved Oxygen Variations at Stations-I and II during Jan–06 Jan–07

Month and Year	*D.O mg/lt*	
	Stn. I - Karwar	*Stn. II - Majali*
Jan-06	6.2	6.5
Feb-06	5	6.2
Mar-06	5.2	6
Apr-06	5.2	5.5
May-06	5.4	5
Jun-06	6	7.3
Jul-06	6.2	7.26
Aug-06	6.2	7
Sep-06	6	6.5
Oct-06	5.4	7
Nov-06	6.1	6.2
Dec-06	6	6.5
Jan-07	6.1	6.5

Similar trend was observed in station II (Majali) with 7.3 mg/l and 7.26 mg/l in the month of June and July 2006 respectively, whereas, low value (5 mg/l) recorded in the month of May 2006 (Table 3.10 and Figure 3.5). Hydrography of Andaman sea have been worked out by Ramesh and Shastry (1976). Subramanya and Ramamohan (1986) studied the variation of sea surface temperature in the Bay of Bengal. Lakshman and Anitha (1986) made a study of physical features of

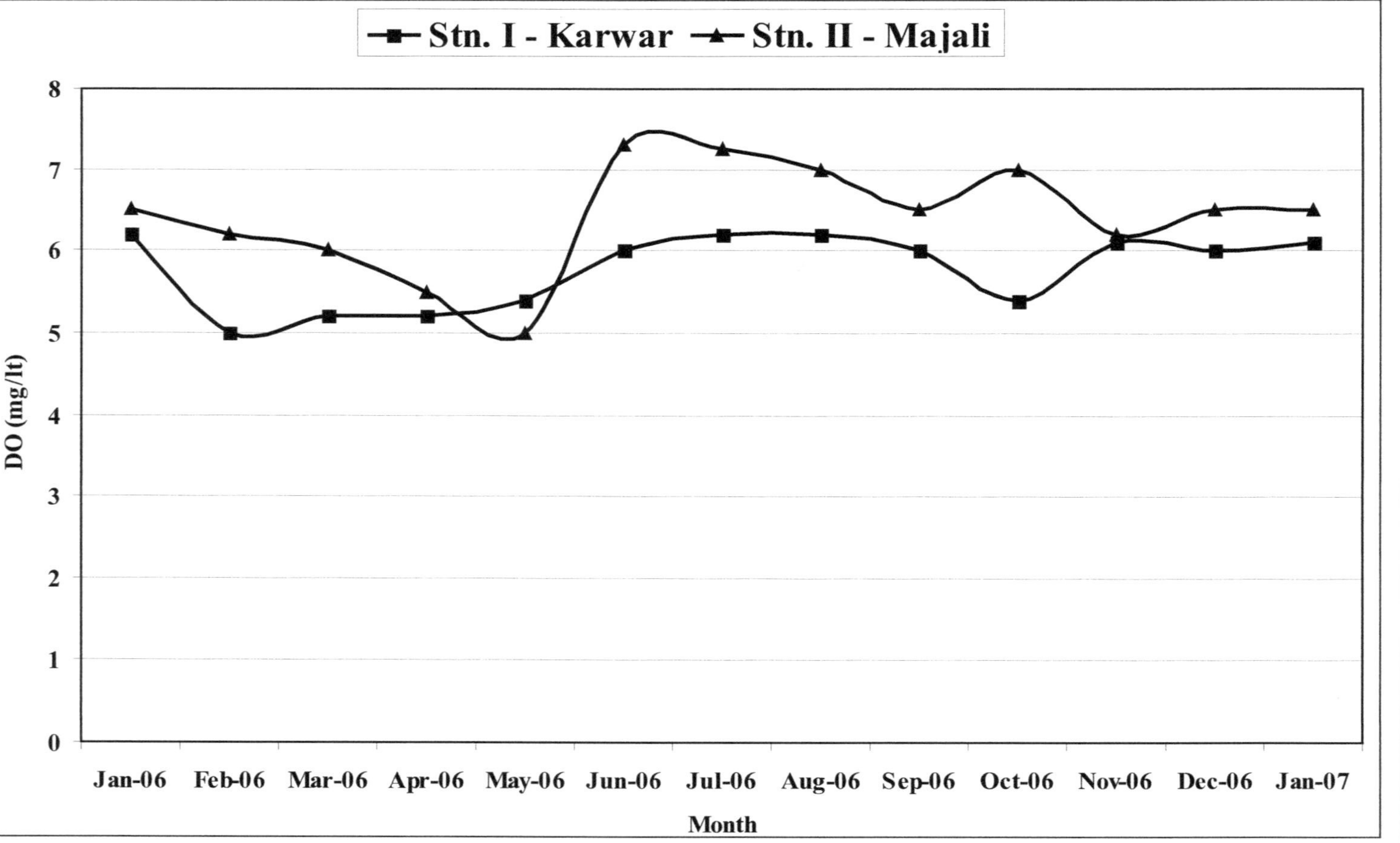

Figure 3.5: Dissolved Oxygen Variations at Stations-I and II during Jan–06 Jan–07.

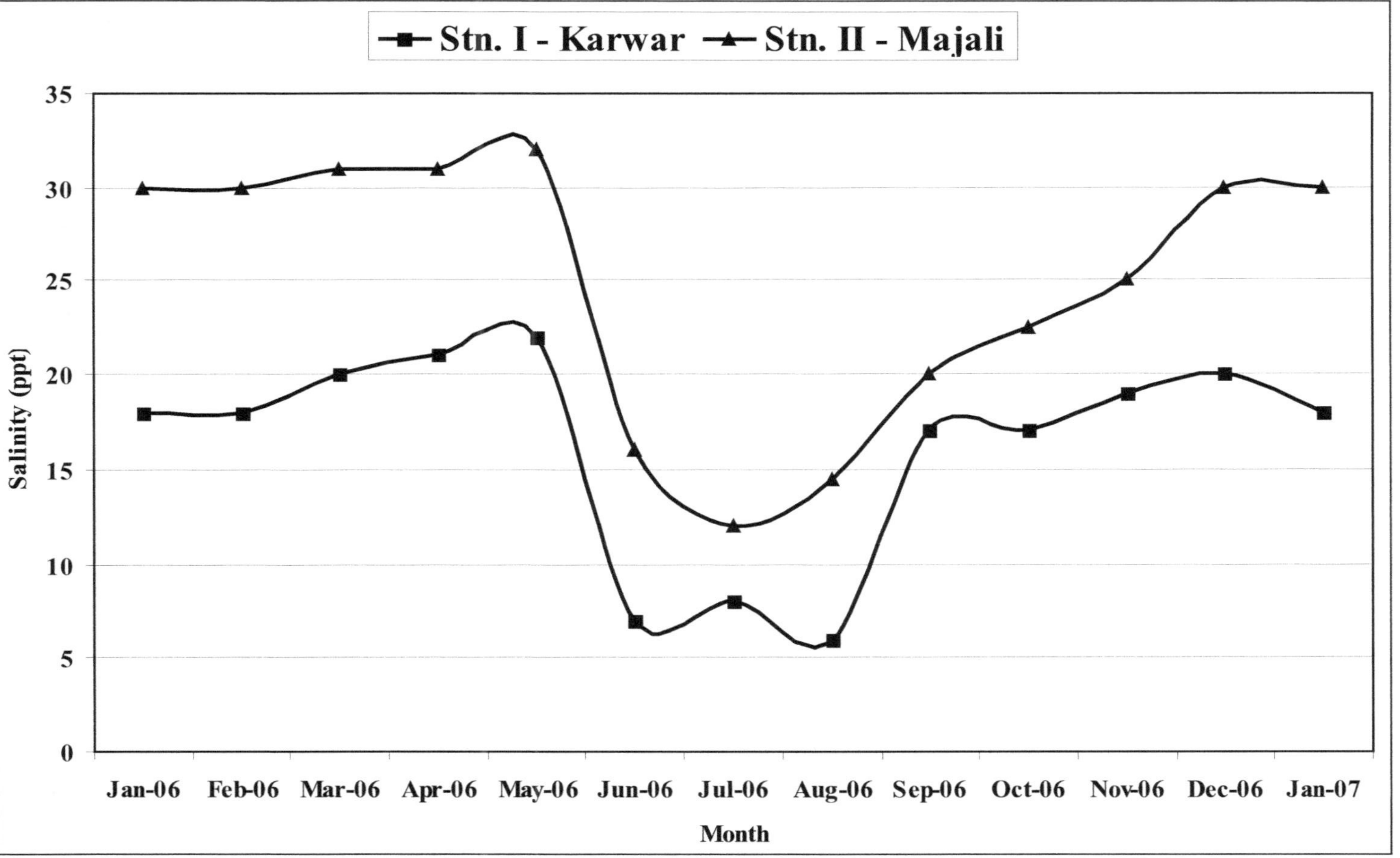

Figure 3.6: Salinity Variations at Stations-I and II during Jan–06 Jan–07.

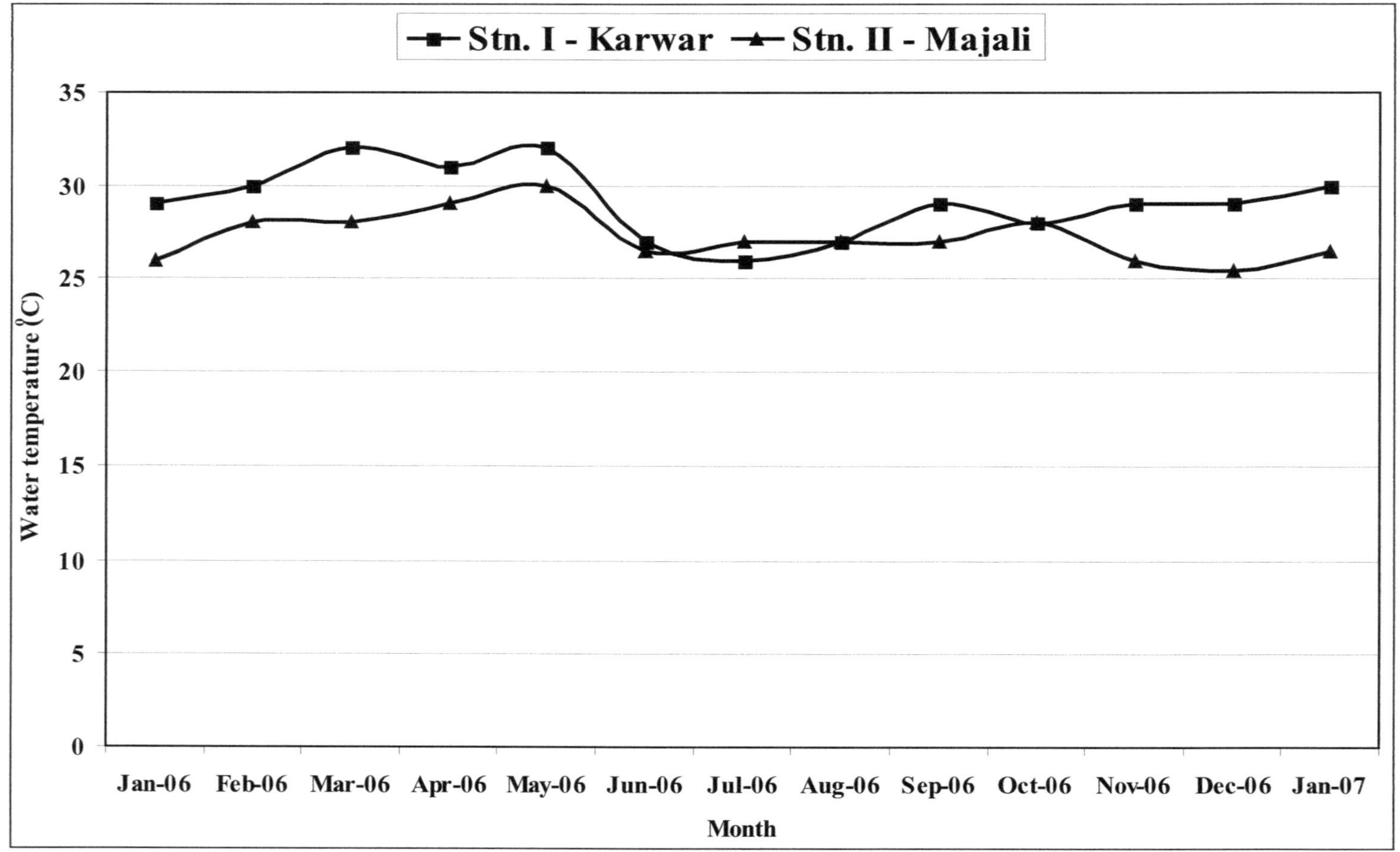

Figure 3.7: Water Temperature Vtariations at Stations-I and II during Jan–06 Jan–07.

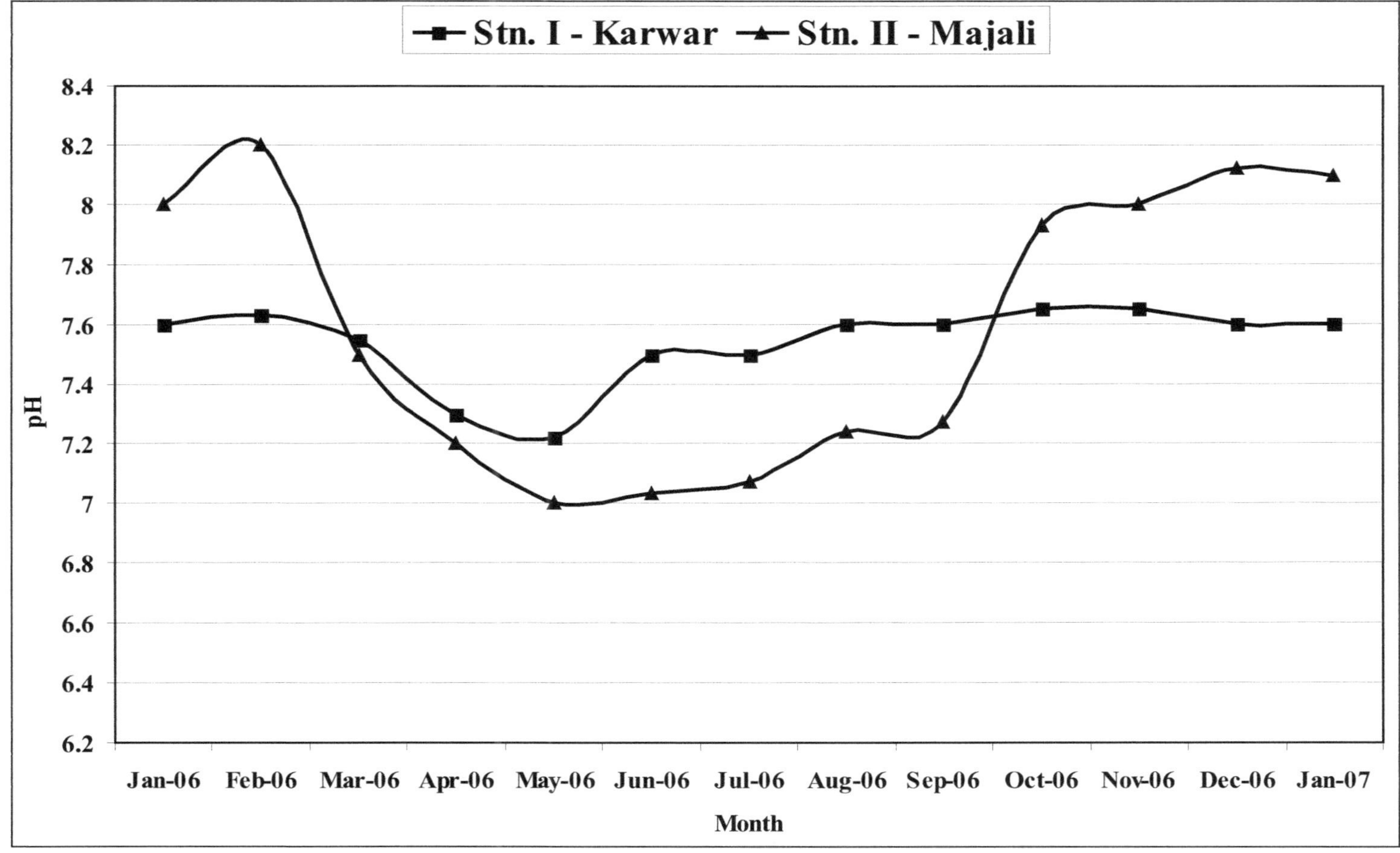

Figure 3.8: pH Variations at Stations-I and II during Jan–06 Jan–07.

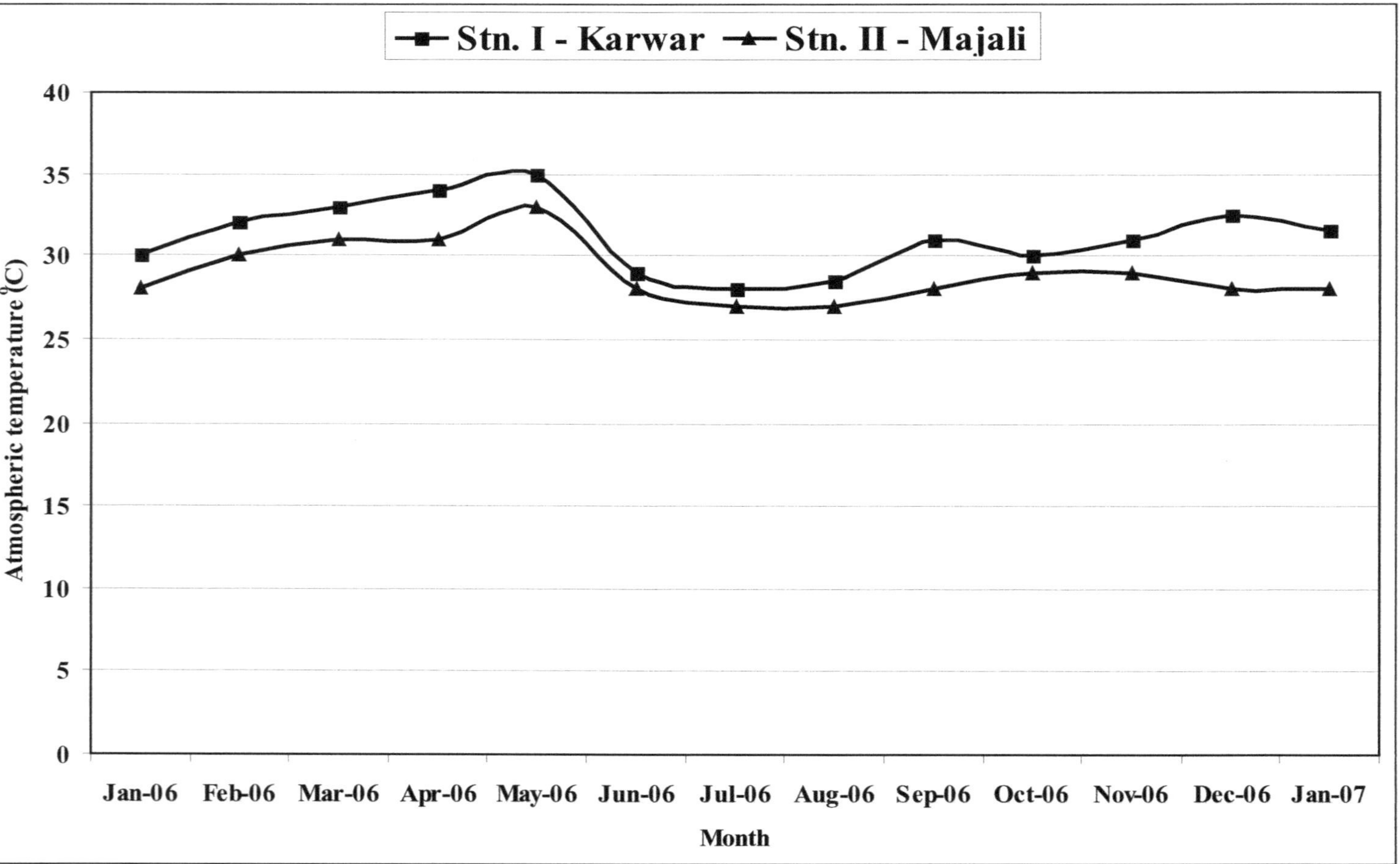

Figure 3.9: Atmospheric Temperature Vtariations at Stations-I and II during Jan–06 Jan–07.

an exposed Sandhu beach at Gopalpur (Orissa). Hydrography of wedge bank in pre-monsoon and monsoon seasons was investigated by Narasimha *et al.* (1979).

The variation in the salinity of the two study stations was undertaken, the highest record salinity was observed at station I is 22 ‰ during the month of May 06 and lowest value of 6 ‰ salinity was recorded.The highest value of salinity at station II is 32‰ during May 06 and lowest record of salinity is 12‰ in the month of July 06. The pioneering works on hydrological studies in the inshore waters of Manglore have been done by (Hariharan, 1986). Hydrological conditions in the fishing grounds of South Kanara have been studied by Benakappa *et al.* (1980) they also observed same kind of readings. Hydrological conditions of Karwar waters was studied by Naik *et al.* (1981). George *et al.* (1958) studied the surface salinity Cochin backwaters with reference to tide.

The monthly variations in the water temperature for a period of 13 months was observed in the station I and II. The highest value of water temperature was observed to be 32°C during the month of May 06 in the station I and lowest record is of 26°C during July 06. The highest values of water temperature at station II is 30°C was noticed during May 06 and lowest 25.5°C is observed during the month of December 06.

Monthly variations in atmospheric temperature was also studied, the results revealed that the highest temperature of 35°C was observed in the month of May 06 in station I (Karwar) and lowest temperature of 28°C was noticed during July 06. A similar trend of highest temperature was recorded in the month of May 2006 is 33°C and lowest in the month of July 06 is 27°C.

Monthly variations in pH of water sample in two study locales was also carried out for a period of 13 months. From the study it is noticed that highest pH 7.65 was noticed during the months of October and November 2006. Lowest record of pH recorded is 7.22 during the month of May 06 in station I (Karwar). Similarly in station II the higher value 8.12 pH was recorded during the month of December 06. Lowest value of pH was noticed during the month of May 06 is 7.0. Generally pH decreases with increase in depth (Venkataswamy Reddy, 1986).

Statistical Analysis

The statistical analysis was carried out by Sigma stat. version. 3.5 statistical package by applying between species and months and among species, later it is presented in a tabular form, for biological parameters catch landings of fishes. The carried statistical analysis are mean, S.D., S.E and correlation. It is observed that in the values, the pairs of variables with positive correlation co-efficient and P- values below 0.050 tend to increase. For pairs with P-values greater than 0.050, there is no significant relationship between the two variables. Similar trend was observed for the 2005 and 20006 (Tables 3.11A-B and 3.12A–B). From the Table 3.13 it was observed that, there was no significant relationship between the two variables for a period of five years ($P > 0.050$).

Table 3.11A: Statistical Analysis of Month-wise Prawn Landings and Species Composition at Station I (Karwar) Fishing Harbour during the Year 2005

Species	Mean	Median	Std. Dev.	Std. Error
P. merguiensis	1900.57	2374	1582.23	598.03
M. affinis	2236.60	2467	1283.72	574.10
M. monoceros	8749.43	4237	9788.76	3699.80
M. dobsoni	23513.86	15670	17334.75	6551.92
P. stylifera	29617.71	33534	18132.37	6853.39
Others	21924.00	24210	14205.75	6353.00

Table 3.11B: Pearson Product Moment Correlation Analysis of Month-wise Prawn Landings and Species Composition at Station I (Karwar) Fishing Harbour during the year 2005

Species	M. affinis	M. monoceros	M. dobsoni	P. stylifera	Others
P. merguiensis	0.792	0.869	0.866	0.631	0.934
M. affinis		0.799	0.810	0.818	0.814
M. monoceros			0.941	0.617	0.970
M. dobsoni				0.713	0.897
P. stylifera					0.664
Others					–

The pair(s) of variables with positive correlation coefficients and P values below 0.050 tend to increase together. For the pairs with negative correlation coefficients and P values below 0.050, one variable tends to decrease while the other increases.

Table 3.12A: Statistical Analysis of Month-wise Prawn Landings and Species Composition at Station I (Karwar) Fishing Harbour during the Year 2006

Species	Mean	Median	Std. Dev.	Std. Error
P.merguiensis	2905.11	1220	3561.86	1187.29
M.affinis	5081.43	5412	3000.67	1134.15
M.monoceros	18158.25	13917	14622.66	5169.89
M.dobsoni	17213.56	18254	11602.51	3867.50
P.stylifera	24126.00	7666	34727.47	15530.60
Others	195.50	195.5	67.18	47.50

For hydrological parameters like D.O., Salinity, Water and Air temperature, the statistical analysis was carried as mean, median. S.D., S.E. One way ANOVA and paired T-test between stations. The difference in the mean values among the treatment groups are greater than would be expected by chance, there is statistically significant difference (P= 0.009) Tables 3.14A and 3.14B this is with reference to air temperature between the stations. The value of pH between the stations was not significant (Table 3.15). For Salinity one way analysis of variance, paired T-test shows the statistically significant difference (P= 0.008) Tables 3.16A to 3.16D. D.O. one way analysis of variance and paired T-test was carried out. The statistical significant difference was observed (P= 0.005) Table 3.17A to 3.17C. For water temperature between two stations one way analysis of variance, paired T- test was calculated. There is statistically significant different was observed (P=0.006) Table 3.18A to 3.18D.

Table 3.12B: Pearson Product Moment Correlation Analysis of Month-wise Prawn Landings and Species Composition at Station I (Karwar) Fishing Harbour during the Year 2006

Species	*M. affinis*	*M. monoceros*	*M. dobsoni*	*P. stylifera*	*Others*
P. merguiensis	0.707	0.941	0.779	0.0629	1
M. affinis		0.619	0.674	–0.473	–1
M. monoceros			0.832	0.209	1
M. dobsoni				0.443	1
P. stylifera					1
Others					–

The pair(s) of variables with positive correlation coefficients and P values below 0.050 tend to increase together. For the pairs with negative correlation coefficients and P values below 0.050, one variable tends to decrease while the other increases.

Table 3.13: Statistical analysis of Species composition of marine fish landings in Uttara Kannada District from 2002 to 2007

Year	*Mean*	*Median*	*Std. Dev.*	*Std. Error*
2002-03	1060.55	594.6	1640.08	294.57
2003-04	889.51	374.4	1608.40	288.88
2004-05	904.55	389.8	1519.99	277.51
2005-06	863.77	455.7	1027.69	190.84
06-07 (Feb-07)	566.68	316.3	717.86	128.93

The pair(s) of variables with positive correlation coefficients and P values below 0.050 tend to increase together. For the pairs with negative correlation coefficients and P values below 0.050, one variable tends to decrease while the other increases.

Table 3.14A: Table Showing Statistical Analysis of Atmospheric temperature

Stations	*Mean*	*Median*	*Std. Dev.*	*Std. Error*
Stn. I - Karwar	31.19	31	2.11	0.58
Stn. II - Majali	29.00	28	1.78	0.49

Table 3.14B: One Way Analysis of Variance

Stations	Mean	Std. Dev.	SEM
Stn. I - Karwar	31.192	2.107	0.584
Stn. II - Majali	29.000	1.780	0.494

The differences in the mean values among the treatment groups are greater than would be expected by chance; there is a statistically significant difference (P = 0.009). Power of performed test with alpha = 0.050: 0.740.All Pairwise Multiple Comparison Procedures (Holm-Sidak method):Overall significance level = 0.05.

Table 3.12C: Paired t-test: Wilcoxon Signed Rank Test

Stations	Median	25 per cent	75 per cent
Stn. I – Karwar	31	29.750	32.625
Stn. II – Majali	28	28.000	30.250

W= -91.000 T+ = 0.000 T-= -91.000.

Z-Statistic (based on positive ranks) = -3.204.

P(est.)= 0.002 P(exact)= <0.001.

The change that occurred with the treatment is greater than would be expected by chance; there is a statistically significant difference (P = <0.001).

Table 3.15: Statistical Analysis of pH

Stations	Mean	Median	Std. Dev.	Std. Error
Stn. I - Karwar	7.54	7.6	0.13	0.04
Stn. II - Majali	7.59	7.5	0.47	0.13

The differences in the median values among the treatment groups are not great enough to exclude the possibility that the difference is due to random sampling variability; there is not a statistically significant difference (P = 0.938).

Table 3.16A: Statistical Analysis of Salinity

Stations	Mean	Median	Std. Dev.	Std. Error
Stn. I - Karwar	16.23	18	5.48	1.52
Stn. II - Majali	24.92	30	7.14	1.98

Table 3.16B: One Way Analysis of Variance
Kruskal-Wallis One Way Analysis of Variance on Ranks

Stations	Median	25 per cent	75 per cent
Stn. I - Karwar	18	14.75	20
Stn. II - Majali	30	19	30.25

H = 7.023 with 1 degrees of freedom. (P = 0.008)

The differences in the median values among the treatment groups are greater than would be expected by chance; there is a statistically significant difference (P = 0.008).

Table 3.16C: Paired t-test:

Stations	*Mean*	*Std. Dev.*	*SEM*
Stn. I - Karwar	16.231	5.48	1.52
Stn. II - Majali	24.923	7.144	1.981
Difference	–8.692	3.099	0.86

t = -10.112 with 12 degrees of freedom. (P = <0.001).

95 percent confidence interval for difference of means: -10.565 to -6.819.

The change that occurred with the treatment is greater than would be expected by chance; there is a statistically significant change (P = <0.001).

Power of performed test with alpha = 0.050: 1.000.

Table 3.16D: Pearson Product Moment Correlation

Stations	*Stn. II - Majali*
Stn. I - Karwar	0.913
Stn. II - Majali	

The pair(s) of variables with positive correlation coefficients and P values below 0.050 tend to increase together. For the pairs with negative correlation coefficients and P values below 0.050, one variable tends to decrease while the other increases.

Table 3.17A: Table Showing Statistical Analysis of DO

Stations	*Mean*	*Median*	*Std. Dev.*	*Std. Error*
Stn. I - Karwar	5.769	6	0.452	0.125
Stn. II - Majali	6.42	6.5	0.665	0.185

Table 3.17B: One Way Analysis of Variance
Kruskal-Wallis One Way Analysis of Variance on Ranks

Stations	*Median*	*25 per cent*	*75 per cent*
Stn. I - Karwar	6	5.35	6.125
Stn. II - Majali	6.5	6.15	7

H = 7.933 with 1 degrees of freedom. (P = 0.005)

The differences in the median values among the treatment groups are greater than would be expected by chance; there is a statistically significant difference (P = 0.005)

Summary

During the present observation 33 verities of fishes belonging to 16 families were noticed at station I. Whereas, in station II there are 21 types of fishes, representing the 15 families were recorded.

Table 3.17C: Paired t-test

Stations	*Mean*	*Std. Dev.*	*SEM*
Stn. I - Karwar	5.769	0.452	0.125
Stn. II - Majali	6.42	0.665	0.185
Difference	–0.651	0.548	0.152

t = -4.285 with 12 degrees of freedom. (P = 0.001).

95 percent confidence interval for difference of means: -0.982 to -0.320.

The change that occurred with the treatment is greater than would be expected by chance; there is a statistically significant change (P = 0.001).

Table 3.18A: Table Showing Statistical Analysis of Water Temperature

Stations	*Mean*	*Median*	*Std. Dev.*	*Std. Error*
Stn. I - Karwar	29.154	29	1.864	0.517
Stn. II - Majali	27.269	27	1.285	0.356

Table 3.18B : One Way Analysis of Variance

Stations	*Mean*	*Std. Dev.*	*SEM*
Stn. I - Karwar	29.154	1.864	0.517
Stn. II - Majali	27.269	1.285	0.356

The differences in the mean values among the treatment groups are greater than would be expected by chance; there is a statistically significant difference (P = 0.006).

Stations	*Diff of Means*	*T*	*Unadjusted P*	*Critical Level*	*Significant?*
Stn. I - Karwar vs. Stn. II - Maajali	1.885	3.002	0.00618	0.05	Yes

Table 3.18C: Paired t-test

Stations	*Mean*	*Std. Dev.*	*SEM*
Stn. I - Karwar	29.154	1.864	0.517
Stn. II - Majali	27.269	1.285	0.356
Difference	1.885	1.57	0.435

t = 4.328 with 12 degrees of freedom. (P = <0.001).

95 percent confidence interval for difference of means: 0.936 to 2.833.

The change that occurred with the treatment is greater than would be expected by chance; there is a statistically significant change (P = <0.001).

A comparative study was carried out in both the study stations in the year 1979 – 80 to 2006-07 from which it can be concluded that station I is more productive than in terms of fish landing than that of station II. More numbers of active fishermen are engaged in the station I (Karwar). In station I almost 319 mechanised boats and 1304 non-mechanised boats are actively involved in fishing activity, which includes, the mechanized gears like Perse seienes, Trawlers, out board motor nets and gill nets.

The station II consisted of is 251 mechanised boats and 1575 non-mechanised boats are engaged in fishing activity. Gill net fishing was the major fishing gear operated in the station II.

An increasing trend of fishery was observed from 1979-80 to 1987-88 in station I, a sudden decline was recorded from 1988-89 to 1996-97, successive decreasing trend recorded in later years.

The station II, showed decreasing trend in fish landing from 1979-80 to 2006-07. Uttaara Kannada district experienced decreasing trend of fish landing over period of two decades *i.e.* from the year 1985-86 to 2006-07 The highest fish catch was recorded (71776.4 MT) value of worth of Rs. 6165.83 lakhs during 1996-97.

The month wise details on the species composition of prawn catch at study station I (Karwar) from the year January 2006-January 2007, consisted of the following species, *Penaeus indicus, P. semisulcatus, Metapenaeus dobsoni, Parapenaeopsis stylifera, P. merguensis* and other miscellaneous groups.

Among prawns landed in station I the composition of *P. stylifera* was 36.55 per cent forming the first dominant species followed by *M. dobsoni,* 29.02 per cent, *M.monoceros* 10.80 per cent, *P. merguensis* 2.35 per cent and *M. affinis* 1.97 per cent and other variety constituted of 19.32 per cent.

In the landing of *M. dobsoni* 32.08 per cent forming the first dominant species followed by *M. monoceros* with 30.08 per cent, *P. stylifera* with 24.98 per cent, *M. affinis* with 7.37 per cent, *P. merguensis* with lowest 5.41 per cent.

In the station I there are 35 verities of fishes were recorded of which squilla formed the major contributor to the fishery with 221 MT of catch over a period of 13 months. Oil sardine formed second important fish (160 MT) and soles formed the third group with 129 MT of catch, *Gerrus* sps. fish formed the least contributor 2 Mt.

In the station II There are 35 verities of fishes were recorded of which oil sardine formed the major contributor to the fishery with 47.5 MT of catch over a period of 13 months. Shell fish formed second important fish (44.3 MT) and scienids formed the third group with 43.5 MT of catch, Lobsters formed the least contributor 1 MT during present work.

During the year 2002-03 to 2007 rays and skates formed the first important fish group with total of 28503.5 tonnes in five years. Followed by prawns with 8665.5 tonnes, mackerel with 8488.8 tonnes, squilla 8430.3 tonnes and least 26.7 tonnes of eels.

The highest values of D.O. recorded in the month of June and July 2006 is 6.2 mg/l in station I and lowest value was observed in the month February 2006 is 5.0 mg/l.

The highest salinity was recorded at station I is 22 ‰ during the month of May and lowest value of 6 ‰ in the month of August. The highest value of salinity at station II is 32‰ during May 06 and lowest record of salinity is 12‰ in the month of July 06.

The highest value of water temperature was observed to be 32°C during the month of May in the station I and lowest record is of 26°C during July The highest value of water temperature at station II is 30°C was noticed during May and lowest 25.5°C is observed during the month of December.

In the month of May highest water temperature of 33°C. was noticed in station I and lowest 28°C during July. A similar trend of highest temperature was recorded in the month of May is 33°C and lowest in the month of July is 27°C.

It is noticed that highest pH 7.65 was noticed during the months of October and November.Lowest record of pH recorded is 7.22 during the month of May in station I. Similarly in station II the higher value 8.12 pH was recorded during the month of December.Lowest value of pH was noticed during the month of May is 7.0.

REFERENCES

Anderson Von Brandt, 1984. Catching methods in fisheries, Fish catching methods of the World. pp. 1-16, Third Edition 1984.

Annigeri. G. G. 1966. Observations on the hydrographical features of the coastal waters of North Kanara during 1956-1966. *Journal of the marine Biological Association of India*.21(1 and 2) Dec, 1979, pp. 128 - 132.

Annigeri. G. G. 1988. Hydrology of the inshore water of Karwar Bay during 1964-66. *Indian Journal of Fisheries*. 15(1 and 2), 1968 pp No. 155-165.

Banerji, S.K. 1958. Fishery survey and statistics Fisheries of the west Coast of India CMFIR, India. pp. 68-73.

Benakappa, S. Reddy, M.P.M. and Hariharan, V., 1980.Hydrolographic conditions in the fishing grounds off Mukkakaup, South Kanara. *Mahasagar Bull. Natn. Inst. Oceanogr.*, 13: 1-7.

Bhat, U.G. 1981. Studies on some Biological aspects and fishery of the fringe scale sardine, *Sardinella fimbriata* from Karwar waters. M.Sc. dissertation.

CMFRI Annual Report of year, 1978-79 and 2005-06.

De Silva, S. S. 1987. Impact of ecotics on the inland fishery resources of Sri Lanka. Furtado. J.I. eds.No. 28, pp. 273-295.

Detelf Quadfasel and Kai.Jancke. 1985. Some graphic observation in North-Western Indian Ocean during the early south-west monsoon 1983. *Mahasagar Bull. Natn. Inst. Oceanogr.* 18(3): 387-394.

Dow, R.L. 1981. Influence of sea temperature cycles on the abundance and availability of marine and estuarine species of commerce. *Marine Dept. Mar. Resour.*2: 775-779.

FAO Fish Identification Key book 1980.

George, M.J., 1958. Sole fisheries, Fisheries of the west coast of India.CMFRI India. pp.51-54.

George, M.J., *et. al* 1963. Observations on the offshore prawn fishery of Cochin.IJF, Vol. -X, No 2,. pp.460-99.

Gunter, G. 1961. Habitat of Juvenile shrimp (family- Penaeidae). *Ecology*. 42(3): 598-600.

Kuttyamma V. J. and Kurian C.V. 1980. Prawn fry resources in the Kayakulam lake. Nat. Symp. Shrimp farming 13: Bay; pp. 49-52.

James, P.S.B.R., 1989. Monsoon fisheries of the west Coast of India prospects, problems and Management. CMFRI Bulletin 45.pp. 251-259.

Jowett, J.G. 1990. Factors related to the distribution and abundance of brown and rainbow trout in New-Zealand clear water rivers. N. 2.J. *Mar. Fresh Wat. Res.*24(3): 429-440.

Kainz, E. Goumann, H.P. 1990. Distribution of some small sized species of Australia running water. *Oester. Fish.*43(11): 265-268.

Kasturirangan, L.R., 1957. A study of the seasonal changes in the dissolved oxygen of surface waters of the sea in the Malabar coast. *Indian J. Fish*, 4(1): 134-149.

Kibria, G. Khaieque, M.A., Rainboth, W.J. 1987. Abundance and distribution of the fin fishes and prawn at five trawling stations of the Meghna River, Bangladesh. Banglades. *J.Zool.* 7(2).

Kurup, N. Surendranath. 1977. On the prawn fishery by trawlers off Purakad South West Coast of India during 1972-76. *Ind.J. Fish.* 33(3): 362.

Kuttyamma V. J. 1975. Studies on the relative abundance and seasonal variation in the occurance of the post-larvae of three species of penaeid prawn in the Cochin Backwaters. *Bull. Dept. of Mar. Sci. Univ. Cochin* 7: 213-219.

Lakshmana Rao, Pattnaik, Anita. 1986. Physical features of an exposed sandy beach at Gopalpur. *Mahasagar Bull. Natn. Inst. Oceanogr.* 19(3): 153-164.

Mhow, Gupta, S.K. Grower 1985. On the fish fauna of Banda District (U.P.). *Int. J. Acad. Ichthyol. Hodinaga*,6(1-2): 97-108.

Manna, B. Goswami, B.C. 1985. Check list of marine and estuarine fishes of Digha West Bengal, India. *Mahasagar,* 18: 489-499.

Marichanny R. and R. Pon Siraimeen, 1979.

Mirza, N.A. Mirza, M.R. 1988. Note on fish fauna of Khanpur Lake, North West, frontiers Province, Pakistan. *Pak. J. Zoo.*20(3): 14-15.

Nair N.B. and Thampy, D.N. 1980. A book of marine ecology publ. The mac-millan Co.

Nagaraj M. and Neelakantan B. 1982. Fish and shell fish seed resources of Kali estuary along with a note on the mariculture potentials in Uttara Kannada. Proc. Symp. Coastal Aquaculture 1: 383-387.

Naik U. G. and B. Neelakantan 1982.Gears and crafts of India, An overview. *J. Indian Fisheries Association,* pp. 245-252.

Narasimha, K.A., G. Sudhakar Rao., Y. Appanna Sastry and W. Venugopalan. 1979. Demersal fishery resources off Kakinada with a note on economics of commercial trawling. IJF, vol. 26, No. 1 and 2, 90-100.

Pradhan, L.B. 1956. Mackerel fishery of Karwar, IJF, vol. III, No.1, pp: 141-185.

Prasad, P.N. Rafiuddin, A.S. Naik, U.G. and Neelakantan B. 1991. Salinity tolerance and oxygen consumption in the juveniles of *Penaeus merguensis* (De man).

Purohit,K.1968.Hydrological features of coastal waters of Cochin. M.Sc. Thesis, School of Marine Science, Cochin.

Quasim, S.Z., 1963.Fisheries Hydrography. *Fish. Technol.* 2(2): 155-157.

Raghupathi Hebbar M. 1995. Marine fishery resources of Karwar and Majali with a special emphasis on *Cynoglossus macrostomus* (Norman).

Ramamurthy. S. 1963. Studies on the hydrological factors in the North Kanara coastal waters. *Indian J. Fish.*, 10 (1A): 75-93.

Ramamurthy, S. and Sukumaran, K.K. 1984. Observation on the prawn fishery of Manglore coast during 1970-1980. *Indian J. Fish.* vol. 31(1): 100-107.

Rajani Nair, 2002. Economically important fin fishes along Kali estuary.

Rajesh, Naik.1994.Studies on the distribution, abundance and composition of fin-fishes along Kali River.

Rajyalakshmi,T.1980.Biology of culturable species of prawn *Penaeus monodon, P. indicus* and *Metapenaeus dodsoni*. Proc. Summ. CIFRI Barrackpore.

Ramaiyan,V., Purushothaman, A.and Natarajan, R. 1990. Checklist of estuarine and marine fishes of Parangipettai coastal water. *Matsya,*12-13: .1-19.

Ramesh, babu and Shaetry, J.S. 1976. Hydrography of the Andaman Sea during late winters. IJMS., 5: 179-189.

Rao, D.P., Sarma, R.V.N., Sastry, J.S. and Premchand, K. 1976. On the lowering of the surface temperature in the Arabian sea with advance of the south-west monsoon.II Tm Symposium on ' Tropical Monsoons' held at Pune 5, Indian, 8-10.

Rathod,J.L.2004.Eco-biology and fishery of the flower tail shrimp *Metapenaeus dobsoni* (Meirs, 1878) along the Karwar waters. Ph.D. Thesis pp.107-121.

Rivonkar, C.V. and Reddy M.P.M.,1989. Studied hydrological conditions off hosbettu, South Kanara coast and its influence on pelagic fisheries. *Indian J. Fish,*36(4): 323-328.

Sengupta, R. Rajagopal, M.D. and Quasim. 1976.Relationship between dissolved oxygen and nutrients in North-Western Indian Ocean. *IJMS.*, 5: 201-211.

Shankarnarayan Y.N. D'Souza, S.N. and Fondekar,S.P.1983.Nitrate mixing in North American Sea. *IJMS*, 12: 181-182.

Shetty. Satish, R. 1984. Seasonal variability of the temperature field off the south west coast of India. *Proc. India. Aca. Sci. (Earth Planet Sci.)* 93: 399-411.

Strickland, J.D.H. and Parsons, T.R.1975. A manual of seawater analysis. Fish. Res. Bd., Canada, Ottawa, p.310.

Sigma Stat. 3.5 Version, Statistical package, as a tool for calculation of mean, mode, Standard Deviation, *etc.*

Subramanyan,R. 1959. Studies on the phytoplankton of the west coast.

Subramanya, I. and Ramohan Rao, S. 1986. Secular variation of sea surface temperature in the Bay of Bengal.*Mahasagar Bull. Natn. Inst. Oceanogr.* 19(3): 165-173.

Sudarshan. D. 1991. Marine Fishery resources in EEZ of India with special reference to deep sea fishing. Sea food export Journal vol. XXIII No. 7 1991 (Aug.).pp.18-24.

Sukumaran, K.K., Deshmukh, V.D. Sudhakara Rao, G., Alagaraja, K. and Sathianandan, T.V., 1993. A study on the prawn fishery of Netracvati- gurupur estuary, Manglore. *Indian J. Fish.* vol. 34(4), pp.382-388.

Sumit Mandal 2003.Biodiversity of prosobranchs along the rocky shore of Majali coast in Uttara Kannada District.

Ulhas G. Naik and B. Neelakantan. 1981.Gears and craft of Karwar. An overview. *J. Indian fisheries Association.* pp.No.245-252.

Chapter 4

Metagenomic Approaches for Ecological Assessment of Fish Gut Microbiomeusing Next Generation Sequencing Techniques

☆ *Kiran D. Rasal and Jitendra K. Sundaray*

ABSTRACT

Advances in DNA sequencing and high performance computing have brought about paradigm progress in the field metagenomics, allowing in depth study of the largely unexplored complex microbial system. The power of genomics analysis is shifted to metagenomics which applied to entire microbes, bypassing the need to isolate and culture individual microbial species. The metagenomics opens the way of study genetic material of microbes directly from environmental samples, thus offers a potential way to access microbial diversity. In case of animals and humans, metagenomicsof gut micro-biota has been studied comprehensively. Although, fishes are very much diverse in terms of varieties and habitat, but only few study regarding fish gut metagenomics has been reported. Ecological assessment, need to identify microbes along with their diversity in aquatic environment as well as in fish gut. In aquaculture sector, feed is the major cost of production. While, the gut microbes are playing important role in feed digestion via microbial fermentation and role in modulation of immune system, which extensively studied in mammals, but very less in case of fishes. Thus, the metagenomic study will be helpful for investigating the microbes from fish gutwhich associate with metabolism as well as immunityand detection of novel microbes which will be helpful for development of probiotics and antibiotics in aquaculture. The present review highlights the metagenomics study in fishes, next generation sequencing platforms along with bioinformatics tools for metagenomic data analysis and future prospectus.

INTRODUCTION

In last decades, next generations sequencing technologies and bioinformatics tools revolutionized the way of study organism at genome wide scale. The metagenomics study provides opportunity to the scientific community to explore the microbes which inhabited in mammals such as human body, oceans, soils, fishes and environment(Frank *et al.*, 2008; Suchodolski, 2011; Handl *et al.*, 2011). Metagenomics is well-defined as the direct analysis of genetic material of genomes from environmental sample (Thomas *et al.*, 2012). Traditional microbial study using culture based methods in laboratory represent very few microbes reported so far from fish gut microbiome. Consequently, large amount of information about microbial diversity in gut of commercially important fishes has been previously missed with traditional microbiology. Therefore, metagenomics studies are getting prominence mainly because of biological significance and availability of bioinformatics tools.

The fish gut or gastrointestinal track is a complex microbial ecosystem which comprises diverse microbial population (Bjorksten, 2006). The gut microbiota of fishes playing role in microbial fermentation of feed as well as maintaining host health (Flint *et al.*, 2008; Round *et al.*, 2009; Wu *et al.*, 2012; and Ringø *et al.*, 2003). Through traditional metagenomics study, diversity of microbes studied based on 16sRNA gene, which is highly diverse among microbes, but it does not provide functional aspect (Vaishampayan *et al.*, 2010). The bacterial diversity has been studied in few species by harboring 16sRNA genes along with NGS platform such as Grass carp, Asian Seabass, Zebrafish, Gibel carp, Atlantic salmo, European sea bass (Han *et al.*, 2010; Roeselers *et al.*, 2011; Wu *et al.*, 2012; Wu *et al.*, 2013, Li *et al.*, 2014; Xia *et al.*, 2014; Zarkasi *et al.*, 2014; Carda-Diéguez *et al.*, 2014). In addition, metagenomics is helpful for identifying the functional gene composition in microbial community based on annotation and database available. Furthermore, metagenomics study supplemented with meta-transcriptomics analysis revealed that expression of microbes at particular time in gut of fish. In case of human gut microbiome, animals and soil microbiome has been well studied (Caporaso *et al.*, 2011; Zeeuwen *et al.*, 2012) and represents various diverse microbes. The bioinformatics tools also revolutionized capability to count and categorize bacterial population. The various algorithms are being available such MG-RAST, IMG/M, MEGAN, CARMA and MetaPhyler for phylogenetic as well as functional annotation. Also many reference databases are available to give functional background to metagenomic datasets such as KEGG, eggnog, COG/KOG, PFAM and TIGRFAM (Thomas *et al.*, 2012).

In fishes, some fish gut microbiome have been studied using next generation sequencing technologies, reported majority of microbes associated with Proteobacteria and phyla of Proteobacteria, Firmicutes, and Bacteroidetes are predominant among bacteria in fish gut microbiome (Xia *et al.*, 2014; Wu *et al.*, 2012; Roeselers *et al.*, 2011). While, in mammals, Firmicutes and Bacteroidetes were the most dominant phyla in gut microbiome (Qin *et al.*, 2010). The majority of bacteria resides in gastrointestinal portions, have profound influence on fish physiology because, microbes play a vital role in digestion. Also, there is change in microbial composition with respect to change in climate. Moreover, this review exemplifies vital information regarding study of fish gut microbiome and will helpful for conducting future research for exploring the microbial diversity in different aspect of study.

Fish Gut-Microbiome

The aquaculture is growing fast with increasing demand of fish for consumption (FAO 2010). Now a day aquaculture facing various problems such as feed cost, diseases, biotic and abiotic stress due climate change. The bacteria are abundant in the gut of fishes which influences fish physiology as well as health (Cahill, 1990). The bacteria are ingested into fish intestine through water and then colonize into the gut to become the resident microbes into fish (Hansen and Olafsen 1999; Ringø *et al.*, 2003, Navarrete, *et al.*, 2010). The composition of gut microbiota varies from fish to fish due their habitat, diet and environmental conditions (Hansen and Olafsen 1999). In fishes gut structures are varies according their food habit. Thus, microbes mainly from the surrounding environment influences the resident fish gut microbial community. The advances in sequencing technology revealed that gut microbiome of fishes dominated by Proteobacteria,Firmicutes, and Bacteroidetes. Among that,many fish species have gut microbiota dominated by dakua-Proteobacteria (Rawls *et al.*, 2004; Kim *et al.*, 2007; Romero and Navarrete 2006; and Ward *et al.*, 2009). Previous study showed that gut microbiota of freshwater fishes generally included of *Aeromonas* sp., *Pseudomonas* sp., *Flavobacterium* species, *Enterobacter* sp., and *Acinetobacter* sp. (Ringø *et al.*, 1995; Ringø and Birkbeck, 1999). *Lactobacillus* spp. is a foremost component of the gastrointestinal micro-biome of most of fishes (Ringo, *et al.*, 1995).

Next Generation Sequencing Techniques for Study Fish Gut Microbiome

The metagenomic study requires various steps from sample collection, DNA extraction, Library preparation according NGS platform and finally data analysis using bioinformatics tools. The sequencing technologies have made great progress over the last decades since the Sanger chain-terminating based technologies. In 2005, new sequencing technology introduced based on principle sequencing by synthesis, 454 sequencing based on pyro-sequencing. This massively parallel sequencing methodology revolutionized the way of genome study. Subsequently, the various NGS platform are being come in market with specific features such as ABI solid, 454 Roche, Illumina, Ion Torrent and Pacific Bio. Among those NGS, 454 roche and Illumina has been extensively used in the field of metagenomic study (Metzker *et al.*, 2010).The sequencing data from NGS instruments has been continues to increase exponentially at the same time decrease of cost of sequencing.

454 GS FLX Titanium

This technology developed in 2005 and commercialized by Roche in 2006, further merged with Titanium chemistry in 2008. This is the first commercially successful NGS platform. The 454 Roche technology has been sequencing read length of 400-700bp with Titanium chemistry. This technology based on principle sequencing by synthesis *i.e.* pyro-sequencing. In 454 sequencing chemistry, single fragment of DNA are ligated with appropriate adaptors which binds to the bead. Further, emPCR(emulsion PCR) is carried out for fragment amplification in water-droplets inclosing one bead with single DNA fragment. After fragment amplification, those beads loaded on optical fiber chip called PicoTiter plate, created from glass fiber bundle which facilitated light guides and other end facing sensitive CCD camera, which detect emitted light. In sequencing run, nucleotides are flowed sequentially in a fixed order across PTP plate. The basic principle pyro-sequencing involves the sequentially addition of all four dNTPs, only those are complementary to template strand are incorporated by DNA polymerase. This reaction releases pyrophosphate which converted via enzyme such as to produce light signal, which detected by CCD camera and passed into sequence of the template. The signal strength is proportional to the number of nucleotides incorporated in a single nucleotide flow. Despite the larger read length, this technology now a days becoming more costly, generates only 500-600mbp data in single run. The 454 pyrosequencing methods have been implemented rapidly by microbiologist for their use in generating detailed sequence based profiles of microbes (Schellenberg, *et al.*, 2009). In last five years, most of culture independent metagenomics study in some fishes undertaken using this technology(Roeselers *et al.*, 2011;Desai *et al.*, 2012, Semova *et al.*, 2012; Wu *et al.*, 2012; Li *et al.*, 2013; Ye *et al.*, 2014).

The Illumina Platform

The Solexa released the Genome Analyzer in 2006 and company then purchased by Illumina in 2007. Then in 2010, Illumina launched HiSeq 2000 which same as Genome analyzer. Most advantage of this technology, coverage and data output, pair end sequencing. Now a day, most of genomic sequencing study going by use of illumina platform (Quail *et al.*, 2012). In this, precise adapters are linked to DNA fragments and then clonally amplification was carried out by emPCR (emulsion PCR) on the surface Ion Sphere Particles. Then, amplified template beads are loaded into wells that are fabricated on a silicon wafer and finally sequencing is performed. During sequencing, each of the four nucleotides is incorporated sequentially and protons are released, where a signal is sensed relative to the number of nucleotide bases incorporated. In order to investigate gut microbial communities in Asian seabass, *Lates calcarifer* and their interactions with hosts during the starvation was studied with this Illumina Hiseq2000 sequencer (Xia *et al.*, 2014).

Ion Torrent

Among the NGS platform, Ion Torrent provides automated pipeline sequencing and with low cost technology compared to other platform. Ion Torrent technology based on semiconductor chip is simpler, faster, more cost effective and scalable than other NGS platform. The workflow of Ion Torrent includes easy to use steps from fully automated library construction and template preparation, to sequence runs on system and finally data analysis using preinstalled Torrent Suites software that runs on the Torrent Server.

To summarize, the NGS technologies in Metagenomics study, limited by cost, low throughput and added modern scientific concepts.

Bioinformatics Tools for Evaluating Metagenomic Data

The development in the bioinformatics algorithm has been helpful for genome data analysis. The various bioinformatics tools are available for metagenomic data analysis depending upon objectives of study. For microbial diversity or phylogenetic study tools used such as 16s RNA -SEED based analysis, RDP, GreenGenes, Silva databases. The functional analysis computational algorithms such as PCATHER, Phylopythia or other similarity based binning software includes such as IMG/M, MG-RAST, MEGAN, CARMA and MetaPhyler along with functional annotation. Among that, IMG/M, MG-RAST, and CARMA are prominently used for metagenomic data analysis.

IMG/M

It consists of metagenomic data integrated with isolated microbial genomes of species from the Integrated Microbial Genomes (IMG) system. It is available at http://img.jgi.doe.gov/m. It mainly used for taxonomic lineage study of microbial population.

MG-RAST

MG-RAST server is afully automated platform for metagenomesanalysis providing quantitative understandings into microbial populations based-on sequence data. (http://metagenomics.anl.gov/). This server is user friendly and freely available, and helpful for analysis of shotgun metagenome data, ITS/16S/18S, and comparison at functional, taxonomic level.

MEGAN

It is also computer program used for large data of metgenome, here set of data compared against databased using BLAST. It computes and discovers the taxonomical content of the metagenome data set using the NCBI site. MEGAN mainly used for phylogenetic analysis and identification of markers (http://ab.inf.uni-tuebingen.de/software/megan).

CARMA

It is a software pipeline for characterizing the taxonomic composition and genetic diversity of short-read metagenomes. It is designed for analysis for metagenome data generated through 454 systems. (http://www.cebitec.uni-bielefeld.de/index.php/technology-platforms/brf).

MetaPhyler

It is used for assessing bacterial composition from Metagenomic data sequences (http://metaphyler.cbcb.umd.edu/). It is a novel taxonomic classifier of shotgun reads for metagenomic which uses phylogenetic marker genes as a taxonomic reference (Liu 2010, 2011). It uses several reference genes and NCBI nr as well as protein database. This is mainly used for short read data such as Illumina.

Future Prospectus

The review indicated that, very less research happen in case of fishes. Further research is required to investigate microbial diversity in commercial important aquaculture fishes. Current research is required in biotic as well as abiotic stress in fishes and co-relation with microbes associated with respective fishes. This could be possible due to availability of high throughput NGS technology which is affordable and giving in depth understanding of genome at functional level along with bioinformatics tools.

Ecological point of view, there is need to know the diversity of microbes in the aquatic environment around the fishes. Because, aquatic environment which influences the gut micro-biota of fishes. Also in aquaculture sector feed cost is major concern of production and also microbes are essential in microbial-fermentation of feed in the gut. Thus, the metagenomic study will be helpful for investigating the microbes from fish gut which associate with metabolism as well as immunity and detection of novel microbes which will be helpful for development of probiotics and antibiotics in aquaculture. Thus, characterizing the commercial important aquaculture fishes using metagenomics approach will be helpful for comparing the different fishes with respect to microbial population and development of appropriate probiotics specific to fishes. Therefore, future research is exciting to understand microbes-host genetic mechanism for development of immunity against diseases and respond to climate change.

Conclusion

The metagenomic analysis is a powerful tool to identify and measure relative proportions of bacterial species in the fish gut over the time in different culture conditions. The present review highlights the metagenomics study in fishes, next generation sequencing platforms along with bioinformatics tools for metagenomic data analysis and future prospectus.

REFERENCES

Björkstén B.2006.The gut microbiota: a complex ecosystem. *ClinExp Allergy* 36: 1215–1217.

Cahill MM. 1990. Bacterial flora of fishes: a review. *Microb. Ecol.*19: 21– 41.

Caporaso, J. G. and others 2011. Global patterns of 16S rRNA diversity at a depth of millions of sequences per sample. *Proceedings of the National Academy of Sciences* 108: 4516-4522.

Carda-Diéguez M1, Mira A, Fouz B.Pyrosequencing survey of intestinal microbiota diversity in cultured sea bass (Dicentrarchuslabrax) fed functional diets. *FEMS Microbiol Ecol.* 2014 Feb. 87(2): 451-9.

Desai, A.R., Links,. MG., Collins, S.A., Mansfoeld, G.S., Drew, M.D., Van Kessel, A.G.and Hill, J.E. 2012. Effects of plant-based diets on the distal gut micro- biome of rainbow trout (*Oncorhyncus mykiss*). *Aquaculture.* 353: 134–142

FAO. 2010. The state of world fisheries and aquaculture 2010. FAO Fisheries and Aquaculture Department, Rome, Italy.

Flint, H.J., Bayer, E.A., Rincon, M.T., Lamed, R. and White, B.A.2008.Polysaccharide utilization by gut bacteria: potential for new insights from genomic analysis. *Nat Rev Microbiol.* 6: 121–131.

Frank, D.N. and Pace, N.R.2008. Gastrointestinal microbiology enters the metagenomics era. *Cur.r Opin. Gastroenterol.*24: 4-10.

Frank, D. N., A. L. S. Amand, R. A. Feldman, E. C. Boedeker, N. Harpaz, and N. R. Pace. 2007. Molecular-phylogenetic characterization of microbial community imbalances in human inflammatory bowel diseases. *Proc. Natl. Acad. Sci.* USA 104: 13780–13785.

Han, S.F., Liu, Y.C., Zhou, Z.G., He, S.X., Cao, Y.A., Shi, P.J., Yao, B.and Ringo, E.2010. Analysis of bacterial diversity in the intestine of grass carp (*Ctenopharyngodon idellus*) based on 16S rDNA gene sequences. *Aquac. Res.* 42(1): 47–56.

Handl, S., S. E. Dowd, J. F. Garcia-Mazcorro, J. M. Steiner, and J. S. Suchodolski. 2011. Massive parallel 16S rRNA gene pyrosequencingreveals highly diverse fecal bacterial and fungal communities in healthy dogs and cats. *FEMS Microbiol.Ecol.*

Hansen, G., and Olafsen, J. 1989. Bacterial colonization of cod (*Gadus morhua* L.) and halibut (*Hippoglossus hippoglossus*) eggs in marine aquaculture. *Appl. Environ. Microbiol.* 55, 1435–1446.

Kim, D. H., Brunt, J., and Austin, B. 2007.Microbial diversity of intestinal contents and mucus in rainbow trout (*Oncorhynchus mykiss*). *J. Appl. Microbiol.*102: 1654–1664.

Lane, D. 1991. 16S/23S *r*RNA sequencing.Nucleic acid techniques in bacterial systematics.

Ley, R.E., Lozupone, C.A., Hamady, M., Knight, R, Gordon, J.I.2008.Worlds within worlds: evolution of the vertebrate gut microbiota. *Nat Rev Microbiol.* 6: 776–788.

Li, X., Yu, Y., Feng W., Yan, Q. and Gong, Y.2012.Host species as a strong determinant of the intestinal microbiota of fish larvae. *J Microbiol.* 50: 29–37.

Li, K., Bihan, M. and Methe, B.A.2013.Analyses of the stability and core taxonomic memberships of the human microbiome. PLoS One 8: e63139.

Li, X., Yan, Q., Xie, S., Hu, W., Yu, Y. and Hu, Z.2013. Gut microbiota contributes to the growth of fast-growing transgenic common carp (*Cyprinus carpio* L.). PLoS ONE 8: e64577.

Liu, Bo, Theodore Gibbons, Mohammad Ghodsi, Todd Treangen and Mihai Pop.2011. Accurate and fast estimation of taxonomic profiles from metagenomic shotgun sequences. *BMC Genomics*.12(Suppl 2): S4.

Liu, Bo, Theodore Gibbons, Mohammadreza Ghodsi, Mihai Pop. MetaPhyler: Taxonomic profiling for metagenomic sequences. *Proceedings of 2010 IEEE Bioinformatics and Biomedicine*: 95-100.

Metzker, M. L. 2010. Sequencing technologies - the next generation. Nat. Rev. Genet. 11, 31–46.

Navarrete, P., Mardones, P., Opazo, R., Espejo, R., and Romero, J. 2010. Oxytetracycline treatment reduces bacterial diversity of intestinal microbiota of atlantic salmon. *J. Aquat. Anim. Health.* 20, 177–183.

Nayak SK. 2010. Role of gastrointestinal microbiota in fish.*Aquac. Res.*41: 1553–1573

Ni J, Yu Y, Zhang T, Gao L.2012.Comparison of intestinal bacterial communities in grass carp, *Ctenopharyngodon idellus*, from two different habitats. *Chin J Oceanol Limnol* 30: 757–765.

Qin, J., Li, R., Raes, J., Arumugam, M., Burgdorf, K. S., Manichanh, C., *et al.*, 2010.A human gut microbial gene catalogue established by metagenomic sequencing. *Nature*. 464, 59–65.

Quail Michael A Miriam Smith, Paul Coupland, Thomas D Otto, Simon R Harris, Thomas R Connor, Anna Bertoni, Harold P Swerdlow and Yong GuA tale of three next generation sequencing platforms: comparison of Ion Torrent, Pacific Biosciences and IlluminaMiSeq sequencers. *BMC Genomics* 2012, 13: 341.

Rawls, J., Samuel, B., and Gordon, J.2004.Gnotobiotic zebrafish reveal evolutionarily conserved responses to the gut microbiota. *Proc. Natl. Acad. Sci. U.S.A.*, 101, 4596–4601.

Ringø E, Sperstad S, Myklebust R, Refstie S, Krogdahl A.2006. Characterisation of the microbiota associated with intestine of Atlantic cod (*Gadus morhua* L.) - The effect of fish meal, standard soybean meal and a bioprocessed soybean meal. *Aquaculture* 261: 829–841.

Ringø, E., Løvmo, L., Kristiansen, M., Bakken, Y., Salinas, I., Myklebust, R., Olsen, R.E. and Mayhew, T.M.2010.Lactic acid bacteria vs. pathogens in the gastrointestinal tract of fish: a review. *Aquacult. Res.*, 41, 451–467.

Ringø, E., Olsen, G.J., Mayhew, T.M. and Myklebust, R. 2003.Electron microscopy of the intestinal microflora of fish. *Aquaculture*, 227, 395–415.

Ringø, E., Strøm, E. and Tabachek, J.A. 1995. Intestinal microfloora of salmonids: a review. *Aquacult. Res.*, 26, 773–789.

Ringø, E. and Birkbeck, T.H. 1999.Intestinal microflora of fish larvae and fry. *Aquacult. Res.*, 30, 773–789.

Roeselers, G., Mittge, E.K., Stephens, W.Z., Parichy, D.M., Cavanaugh, C.M.,Guillemin, K.and Rawls,J.F. 2011. Evidence for a core microbiota in the zebra fish. *ISME J.* 5: 1595–1608.

Romero, J., and Navarrete, P.2006. 16S rDNA-based analysis of dominant bacterial populations associated with early life stages of coho salmon (*Oncorhynchus kisutch*). *Microb. Ecol.* 38, 1–26.

Round JL, Mazmanian SK.2009. The gut microbiota shapes intestinal immune responses during health and disease. *Nat Rev Immunol.* 9: 313–323.

Schellenberg J, Links MG, Hill JE, Dumonceaux TJ, Peters GA, Tyler S, Ball TB, Severini A, Plummer FA. 2009.Pyrosequencing of the chaperonin-60 universal target as a tool for determining microbial community composition. *Appl Environ Microbiol*.75(9): 2889-98.

Semova I, Carten JD, Stombaugh J, Mackey LC, Knight R, Farber SA, Rawls JF. 2012. Microbiotaregulate intestinal absorption and metabolism of fatty acids in the zebrafish. *Cell Host Microb*. 12: 277–288.

Suchodolski, S. COMPANION ANIMALS SYMPOSIUM: Microbes and gastrointestinal health ofdogs and cats.*J. Ani. Sci*. 2011, 89: 1520-1530.

Thomas T, Gilbert J, Meyer F.2012. Metagenomics - a guide from sampling to data analysis. *Microb Inform Exp*. 2(1): 3.

Vaishampayan PA, Kuehl JV, Froula JL, Morgan JL, Ochman H, Francino MP.2010. Comparative metagenomics and population dynamics of the gut microbiota in mother and infant. *Genome Biol Evol*. 2: 53–66.

Ward,N., Steven, B., Penn, K., Methé, B., and Detrich, W. III.2009. Characterization of the intestinal microbiota of two Antarctic notothenioid fish species. Extremophiles. 13: 79–685.

Wu S, Wang G, Angert ER, Wang W, Li W, Zou H.2012. Composition, diversity, and origin of the bacterial community in grass carp intestine. PLoS One 7: e30440.

Wu S, Wang G, Angert ER, Wang W, Li W, Zou H.2012. Composition, diversity, and origin of the bacterial community in grass carp intestine. Plos One.7(2): e30440.

Wu SG, Gao TH, Zheng YZ, Wang WW, Cheng YY, *et al*.2010.Microbial diversity of intestinal contents and mucus in yellow catfish (*Pelteobagrus fulvidraco*). *Aquaculture*. 303: 1–7.

Xia Jun Hong, Grace Lin, Gui Hong Fu, Zi Yi Wan, May Lee, Le Wang, Xiao Jun Liu and Gen HuaYue.2014. The intestinal microbiome of fish under starvation. *BMC Genomics*. 15: 266.

Ye, L., Amberg, J., Chapman, D.,Gaikowski, M., and Liu, W.T. 2014. Fish gut microbiota analysis differentiates physiology and behavior of invasive Asian carp and indigenous American fish.*ISME J*.8: 541–551.

Zarkasi KZ1, Abell GC, Taylor RS, Neuman C, Hatje E, Tamplin ML, Katouli M, Bowman JP.2014.Pyrosequencing-based characterization of gastrointestinal bacteria of Atlantic salmon (*Salmo salar* L.) within a commercial mariculture system. *J. Appl.Microbiol*. 117(1): 18-27.

Zeeuwen P, Boekhorst J, van den Bogaard EH, de Koning HD, van de Kerkh of PMC, Saulnier DM, van Swam II, van Hijum S, Kleerebezem M, Schalkwijk J, Timmerman HM.2012. Microbiome dynamics of human epidermis following skin barrier disruption. *Genome Biol*.13(11): R101.

Chapter 5

Bioactive Peptides from Fish Processing Waste: A Review

☆ *G.G. Phadke, F.R. Sofi, A.U. Pagarkar, H. Singh and S.B. Satam*

ABSTRACT

Fish processing industries generates large amount of waste in the form of head, skin, fishframes, viscera which are generally discarded or utilized for preparation of low value products such as fish meal. Bioactive peptides generated from hydrolysis of proteins are the focus of current research. Waste generated during fish processing can be used as a source of raw material for hydrolysis of protein to yield highly valued bioactive peptides. These biologically active peptides exert various physiological functions such as antioxidant activity, antihypertensive activity, antimicrobial activity, anticoagulant activity, anticancer activity and many more. Production of bioactive peptides from fish waste will be beneficial for consumers as well as processors.

Keywords: *Fish waste, Bioactive peptides, Antihypertensive, Antioxidant, Antimicrobial, Anticoagulant.*

INTRODUCTION

Marine fisheries of the country is contributed 3.59 million tonnes in 2014 and more than 70 per cent of this production has been utilized for processing. As a result, every year a considerable amount of processing waste is discarded, which include fish, skin, heads, bones and viscera. It is estimates that fish waste produced 23 per cent of fish production. Hence it is found that there is a great potential in fish processing industry to convert and utilize fish waste as valuable by products. However, fish waste products such as fish oil, fish meal and fish silage *etc.*, have low economic value. Recently technology is developed for production of bioactive compounds from fish waste for human consumption. Furthermore, it is also observed that some of the bioactive compounds have possesses nutracecuticals in human health.

Currently, efforts are being made for utilization of fish processing waste for preparation of high market value products like fish protein hydrolysates (FPH). The proteins in the fish waste undergo degradation leading to formation of low molecular weight peptides rich in bioactive properties (Slizyte*et al.*, 2005a; Nalinanon*et al.*, 2011; Pires*et al.*, 2012).Keeping in view,an importance of bioactive properties of hydrolysates from fish waste has been reviewed in this paper.

Fish Waste as a Source of Protein

Fish is considered as a rich source of proteins due to their high nutritional value and their ability to coordinate growth and other metabolic functions in living organisms. The post harvest losses coupled with generation of processing waste will amount to a total loss of 50-60 per cent of the raw material which is quite huge (FAO, 1998). A large volume of solid waste is processed into fish meal to be used as animal feed. At times, the fish processing waste is disposed into open waters causing environmental pollution (Islam *et al.*, 2004; Pagarkar *et al.*, 2006). Considerable amounts of fish processing waste generated currently impose a cost burden on the seafood industry in terms of waste disposal, with little benefit generated. Sustainable use of fish processing wastes and low value fish for the preparation of hydrolysates which are rich in bioactive peptides is a viable option. The assessment of their bioactivities, functional properties and application in different product formulation is challenging and calls for deeper research inputs (He *et al.*, 2004). The utilization of the fish processing wastes for the preparation of hydrolysates is economically a better option.

Protein Hydrolysate from Fish Waste

The fish protein hydrolysatescan be defined as proteins that are chemically or enzymatically cleaved to peptides of varying mass (Rustad, 2003). Application of proteases to get FPH is an efficient process, wherein, peptides possessing bioactive properties are obtained from fish processing waste and underutilized fish (Slizyte *et al.*, 2005b; Nalinanon *et al.*, 2011; Pires *et al.*, 2012). The hydrolysis of proteins of high molecular weight will lead to the formation of protein hydrolysates comprising mixture of peptides (Shirai and Ramirez, 2011).

Bioactive Peptides from Hydrolysis of Fish Waste

Peptides possessing biological or nutritional properties which improve the functioning of an organism's body as well as which enhance food quality are referred to as 'bioactive peptides' (Hartmann and Meisel, 2007). These are short sequences of amino acids which are inactive within the sequence of parent protein and may be released by proteolytic hydrolysis using commercially available enzymes or proteolytic microorganisms (Vercruysse*et al.*, 2005). Bioactive peptides may be readily present in foods or protein sources as unbound compounds or they may only be released into their active forms upon cleavage from a native protein source. The release of functional and potent bioactive peptides from intact proteins in various studies have been achieved by various methods such as fermentation (Cho *et al.*, 2005), enzymatic hydrolysis (Je *et al.*, 2007) and reactions with acid or alkaline solvents (Solina*et al.*, 2007). Beneficial effects of bioactive peptides derived from food proteins and associated human physiological system are presented in Figure 5.1.

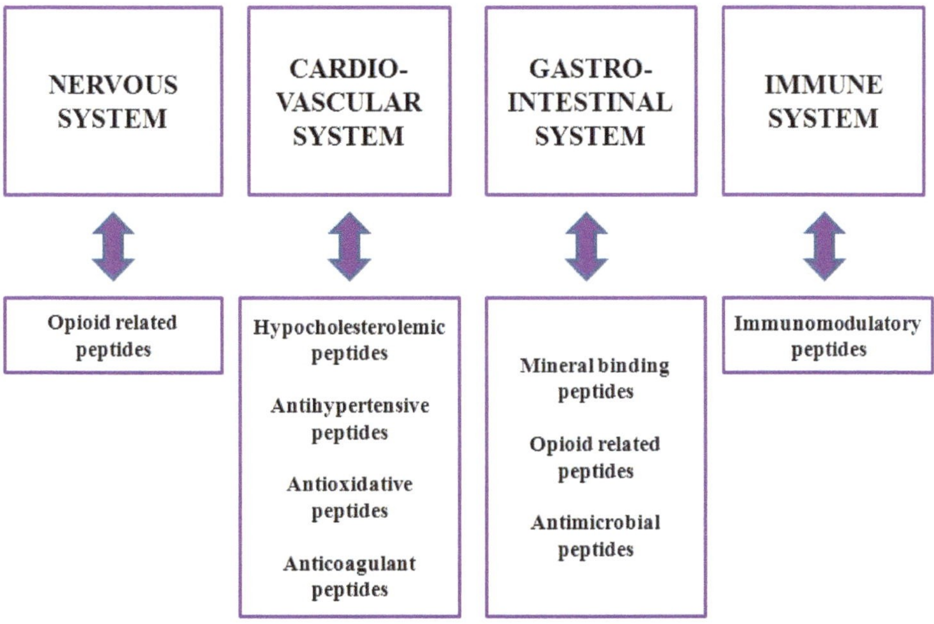

Figure 5.1: Beneficial Effects of Bioactive Peptides in Physiological Systems.

Bioactive Properties of Fish Waste Protein Hydrolysates

Proteins from fish sources and their hydrolysis products and individual peptides are documented to exhibit bioactive properties. These properties include; antihypertensive, antioxidant (Raghavan and Kristinsson, 2008; Raghavan*et al.*, 2008), antithrombotic, anticoagulant, anticancer and antimicrobial (Vercruysse*et al.*, 2005). Some peptides are known to reduce low density lipoprotein (LDL)-cholesterol levels and preventing atherosclerosis and other cardiovascular diseases (Yoshikawa and Chiba, 1992; Turpeinen *et al.*, 2009).

Antihypertensive Activity

Hypertension affects about a quarter of the world's population and is a major, yet controllable risk factor in cardiovascular disease and related complications, the biggest cause of death (Wijesekara and Kim, 2010). Currently, antihypertensive medications such as captopril and analapril, marketed under the trade names Accupril, Altace, Capoten, Lotensin, Monoril, Prinvil, Vasotec and Zestril stabilize blood pressure without removing the root cause, which is as yet unknown (Ahhmed and Muguruma, 2010). Although these synthetic ACE inhibitory drugs have demonstrated their usefulness, they are not entirely without side effects, such as cough, loss of taste, renal impairment and angioneurotic edema (Cooper *et al.*,

2006). Therefore, search for natural safe and more effective ACE-inhibitory agents as alternative ones is necessary for the prevention and treatment of hypertension.

ACE inhibitory peptides have been isolated and characterized from a number of fish sources, either from the muscle proteins or from waste and discards from fish processing. Alaska pollock (*Theragra chalcogramma*) skin gelatin hydrolysates produced using alcalase, pronase E and collagenase demonstrated a high ACE inhibitory activity. Two peptides, Gly-Pro-Leu and Gly-Pro-Met were isolated which were showing high ACE inhibition (Byun and Kim, 2001).A novel ACE inhibitory peptide with sequence Phe-Gly-Ala-Ser-Thr-Arg-Gly-Ala and IC_{50} value of 14.7 μM from Alaska pollock frame protein hydrolyzed with pepsin by ultrafiltration and consecutive gel filtration, ion–exchange and HPLC was isolated (Je *et al.*, 2004).

Antioxidant Activity

Lipid oxidation is of great concern to the food industry and consumers because it leads to the development of undesirable off-flavors, off-odors, dark colors, deterioration of taste and formation of potentially toxic reaction products (Noguchi and Niki, 1999).Furthermore cancer, coronary heart disease and Alzheimer's diseases are also reported to be caused in part by oxidation or free radical reactions in the body (Diaz *et al.*,1997). Inhibition of lipid peroxidation is important in order to prevent foods from undergoing deterioration and to provide protection against oxidation mediated diseases. Antioxidants are substances used to prolong the shelf life and maintain the nutritional quality of lipid-containing foods and to modulate the consequences of oxidative damage in the human body. Many synthetic antioxidants such as butylated hydroxy toluene (BHT), butylatedhydroxy anisole (BHA), tertiary butyl hydro quinone (TBHQ) and propyl gallate (PG) are used in the food and pharmaceutical industries to retard lipid oxidation (Bernardini*et al.*, 2011). However, the use of these synthetic antioxidants must be strictly controlled due to potential health issues (Bougatef*et al.*, 2010). Hence, it is necessary to identify alternative natural, safe sources of food antioxidants especially natural antioxidants (Bernardini *et al.*, 2011). There is growing interest in the replacement of synthetic antioxidants with natural antioxidants from food sources for their potential health benefits with little or no side effects (Bahareh and Amin, 2010). Natural antioxidants are claimed to strengthen the body's defenses through dietary supplementation and inhibit lipid oxidation in foods (Je *et al.*, 2007; Jung *et al.*, 2007b).

Peptides derived from enzymatic hydrolysis of various fish proteins are reported to possess antioxidant activity based on the nature, size and composition of the different peptide fractions and the protease specificity (Klompong*et al.*, 2007). Yellowfin sole frame protein hydrolysates' (YFPHs) fractionation resulted in an antioxidant peptide composed of 10 N-terminal amino acid residues, Arg-Pro-Ser-Phe-Ser-Leu-Glu-Pro-Pro-Tyr, with molecular mass of 13 kDa (Jun *et al.*, 2004). Alaska pollock frame protein hydrolyzed with mackerel intestine crude enzyme (MICE) gave a peptide Leu-Pro-His-Ser-Gly-Tyr with molecular weight 672 Da, which demonstrated hydroxyl radical scavenging activity (Je *et al.*, 2005c). Yellow stripe trevally (*Selaroides leptolepis*) muscle, hydrolyzed by Alcalase exhibited antioxidant activity through the radical scavenging, reducing and metal chelating mode of lipid oxidation (Klompong *et al.*, 2007).

Anticoagulant Activity

Plasma serine proteases known as blood clotting factors are involved in blood coagulation pathway. These factors interact with calcium and phospholipid surface to produce a tough fibrin meshwork, which reinforces the platelet plug and stops bleeding until tissue repair can occur (Davie *et al.*, 1991). Enzymatically hydrolyzed fish muscle peptides have also shown anticoagulant and antiplatelet properties tested *in vitro* and these results have suggested the capability of fish peptides to inhibit coagulation factors in the intrinsic pathway of coagulation (Rajapakse*et al.*, 2005). An anticoagulant peptide purified from yellowfin sole FPH inhibited the activated coagulation factor XII by forming an inactive complex regardless of Zn^{2+}(Rajapakse*et al.*, 2005). Jung *et al.* (2007a) purified a novel protein against blood coagulation factor Va from granulated ark (marine bivalve). An anticoagulant oligopeptide isolated from blue mussel, *Mytilus edulis* with 2.5 kDa molecular mass prolonged thrombin and activated thromboplastin time (Jung and Kim, 2009). Prolongation of thrombin time and activated partial thromboplastin time was also reported for goby muscle protein hydrolysates produced with bacterial alkaline proteases (Nasri*et al.*, 2012).

Antimicrobial Activity

Antimicrobials in food are used for inhibiting the growth of food spoilage microbes and their mode of action has been by altering cell membrane permeability to substrates and by creating intolerable pH conditions for microbial growth (Back *et al.*, 2009). Antimicrobial peptides have been reported to exhibit a broad range of activities against various microorganisms including Gram-positive and Gram-negative bacteria, virus, fungi, and protozoa (Bernardini *et al.*, 2011).

Piscidins (antimicrobial peptides) isolated from striped bass have shown broad spectrum antimicrobial activity and have proved potent against bacteria that have shown resistance to antibiotics (Chekmenev *et al.*, 2006). Antibacterial activity was reported against gram positive and gram negative bacteria in peptide fractions of enzymatically hydrolyzed snow crab byproducts(Beaulieu *et al.*, 2010). Cysteine-rich antibacterial peptide inhibiting the growth of bacteria such as *Escherichia coli, Bacillus subtilis* and *Staphylococcus aureus* and antimicrobial peptides from muscle protein sources (Bernardini *et al.*, 2011) was reported. Kenojeinin-I antimicrobial peptide isolated from fermented skate skin was effective in inhibiting the growth of *B. subtilis, E. coli* and *Saccharomyces cerevisiae* (Cho *et al.*, 2005).Antibacterial activity of half-fin anchovy pepsin hydrolysate against the growth of *Escherichia coli* CGMCC 1.1100, *Pseudomonas fluorescens* CICC 20225, *Proteus vulgaris* CICC 20049 and *Bacillus megaterium* CICC 10324 was reported by Song *et al.* (2011) with the minimal inhibitory concentration values ranging from 28.38 to 56.75 µg/ml.

Conclusion

There are various bioactive peptides generated from enzymatic hydrolysis of fish or fish waste possessing various bioactivities other than those described above such as anticancer activity, anti-inflammatory activity. Currently researchers are working to identify the peptides responsible for particular bioactive property. These bioactive peptides generated from enzymatic hydrolysis of fish waste will be beneficial for consumers as well as processors. Therefore, it will be a great challenge and opportunity for research on utilization of fish processing waste in India.

REFERENCES

Ahhmed, A.M. and Muguruma, M., 2010.A review of meat protein hydrolysates and hypertension. *Meat Sci.*, 86: 110-118.

Back, S.Y., Jin, H.H. and Lee, S.Y., 2009. Inhibitory effect of organic acids against *Enterobacter sakazakii* in laboratory media and liquid foods. *FoodControl,* 20(10): 867-872.

Bahareh, H.S. and Amin, I., 2010. Antioxidative peptides from food proteins: a review. *Peptides,* 31: 1949-1956.

Beaulieu, L., Thibodeau, J., Desbiens, M., Saint-Louis, R., Zatylny-Gaudin, C. and Thibault, S., 2010. Evidence of antibacterial activities in peptide fractions originating from snow crab (*Chionoecetes opilio*) by-products.*Probiotics and Antimicrobial Proteins,* 2(3): 1-13.

Bernardini, R.D., Harnedy, P., Declan-Bolton, D., Kerry, J., O'Neill, O. and Mullen, E., 2011. Antioxidant and antimicrobial peptidichydrolysates from muscle protein sources and by-products: review. *Food Chem.,* 124: 1296-1307.

Bougatef, A., Nedjar-Arroume, N., Manni, L., Ravallec, R., Barkia, A., Guillochon, D. and Nasri, M., 2010.Purification and identification of novel antioxidant peptides from enzymatic hydrolysates of sardinella (*Sardinella aurita*) by-products proteins. *Food Chem.,* 118(3): 559-565.

Byun, H.G. and Kim, S.K., 2001. Purification and characterisation of angiotensin I converting enzyme (ACE) inhibitory peptides from Alaska pollock (*Theragra chalcogramma*) skin. *Proc. Biochem.,* 36(12): 11551162.

Chekmenev, E.Y., Vollmar, B.S., Forseth, K.T., Manion, M.N., Jones, S.M., Wanger, T.J. and Cotten, M., 2006.Investigating molecular recognition and biological function at interfaces using piscidins, antimicrobial peptides from fish. *BiochimicaetBiophysicaActa (BBA) - Biomembranes,* 1758(9): 1359-1372.

Cho, S.H., Lee, B.D., An, H. and Eun, J.B., 2005.Kenojeinin I, antimicrobial peptide isolated from the skin of the fermented skate, *Raja kenojei. Peptides,* 26(4): 581-587.

Cooper, W.O., Hernandez-Diaz, S., Arbogast, P.G., Dudley, J.A., Dyer.S., Gideon, P.S., Hall,K. and Ray, W.A., 2006.Major congenital malformations after first-trimester exposure to ACE inhibitors.*N. Engl. J. Med.,* 354: 2443-2451.

Davie, E.W., Fujikawa, K. and Kisel, W., 1991. The coagulation cascade: initiation, maintenance and regulation. *Biochem.,*30: 10363-10370.

Diaz, M.N., Frei, B., Vita, J.A. and Keaney, J.F., 1997. Antioxidantsand atherosclerotic heart disease.*N. Engl. J. Med.*, 337: 408-416.

FAO, 1998.Aquaculture production statistics. FAO Fisheries Circular no. 815, Rev. 10, Food and Agriculture Organization of the United Nations, Rome.

Hartmann, R. and MEISEL, H., 2007. Food-derived peptides with biological activity: From research to food applications. *Curr.Opin.Biotechnol.*18: 163-169.

He, K., Song, Y., Daviglus, M.L., Liu, K., Van-Horn, L., Dyer, A.R., Goldbourt, U. and Greenland, P., 2004.Fish consumption and incidence of stroke - A meta-analysis of cohort studies.*Stroke,* 35: 1538-1542.

Islam, M.D., Khan, S. and Tanaka, M., 2004. Waste loading in shrimp and fish processing effluents: potential source of hazards to the coastal and nearshore environments. *Marine Poll. Bull.*, 49: 103-110.

Je, J.Y., Oian, Z.J., Byun, H.G. and Kim, S.K., 2007. Purification and characterization of an antioxidant peptide obtained from tuna backbone protein by enzymatic hydrolysis. *Process Biochem.*,42(5): 840 846.

Je, J.Y., Park, P.J. and Kim, S.K., 2005. Antioxidant activity of a peptideisolated from Alaska pollock (*Theragra chalcogramma*) frame protein hydrolysate. *Food Res. Intl.*,38: 45-50.

Je, J.Y., Park, P.J., Kwon, J.Y. and Kim, S.K., 2004. Novel angiotensin I converting enzyme inhibitory peptide from Alaska pollock (*Theragra chalcogramma*) frame protein hydrolysate. *J. Agric. Food Chem.*, 52(26): 78427845.

Jun, S.Y., Park, P.J., Jung, W.K. and Kim, S.K., 2004. Purification and characterization of an antioxidative peptide from enzymatic hydrolysate of yellowfin sole (*Limanda aspera*) frame protein. *Eur. Food Res. Technol.*, 219: 20-26.

Jung, W.K. and Kim, S.K., 2009. Isolation and characterization of an anticoagulant oligopeptide from blue mussel, *Mytilus edulis*. *Food Chem.*, 117(4): 687-692.

Jung, W.K., Jo, H., Qian, Z., Jeong, Y., Park, S., Choi, I. and Kim, S., 2007a. A novel anticoagulant protein with high affinity to blood coagulation factor Va from *Tegillarca granosa*. *J. Biochem. Mol. Biol.*, 40(5): 832-838.

Jung, W.K., Oian, Z.J., Lee, S.H., Choi, S.Y., Sung, N.J., Byun, H.G. and Kim, S.K., 2007b. Free radical scavenging activity of a novel antioxidative peptide isolated from *in vitro* gastrointestinal digests of *Mytilus coruscus*. *J. Med. Food,* 10(1): 197202.

Klompong, V., Benjakul, S., Kantachote, D. and Shahidi, F., 2007.Antioxidative activity and functional properties of protein hydrolysate of yellow stripe trevally (*Selaroides leptolepis*) as influenced by the degree of hydrolysis and enzyme type. *Food Chem.*, 102: 1317-1327.

Nalinanon, S., Benjakul, S., Kishimura, H. and Shahidi, F., 2011. Functionalities and antioxidant properties of protein hydrolysates from the muscle of ornate threadfin bream treated with pepsin from skipjack tuna. *Food Chem.*, 124: 1354-1362.

Nasri, R., Amor, I.B., Bougatef, A., Nedjar-Arroume, N., Dhulster, P., Gargouri, J., Chaabouni, M.K. and Nasri, M., 2012. Anticoagulant activities of goby muscle protein hydrolysates. *Food Chem.*, 133: 835-841.

Noguchi, N. and Niki, E., 1999. Chemistry of active oxygen species and antioxidants. *In*: Antioxidant Status, Diet, Nutrition, and Health,Edt. Papas, A. M., Boca Raton: CRC Press,Boca Raton, USA, pp. 3-20.

Pagarkar, A. U., Basu, S. and Mitra, A., 2006 Preparation of bio-fermented and acid silage from fish waste and its biochemical characteristics. *Asian Jr. of Microbiol. Biotech. and Enviro. Sci.*, 8 (2) 381-387.

Pires, C., Clemente, T. and Batista, I., 2012.Functional and antioxidative properties of protein hydrolysates from Cape hake by-products prepared by three different methodologies.*J. Sci. Food Agric.*, 93: 771-780.

Raghavan, S. and Kristinsson, H.G., 2008.Antioxidative efficacy of alkali-treated tilapia protein hydrolysates: a comparative study of five enzymes. *J. Agric. Food Chem.*, 56: 1434-1441.

Raghavan, S., Kristinsson, H.G. and Leeuwenburgh, C., 2008.Radical scavenging and reducing ability of tilapia (*Oreochromis niloticus*) protein hydrolysates.*J. Agric. Food Chem.*, 56: 10359–10367.

Rajapakse, N., Jung, W., Mendis, E., Moon, S. and Kim, S.K., 2005. A novel anticoagulant purified from fish protein hydrolysate inhibits factor XIIa and platelet aggregation. *Life Sci.*, 76: 2607-2619.

Rustad, T., 2003.Utilisation of marine by-products.*Electronic J. Envi. Agric. Food Chem.*, 2: 458-463.

Shirai, K. and Ramirez, J.C., 2011. Utilization of fish processing by-products for bioactive compounds.*In*: Fish Processing - Sustainability and New Opportunities, Edt. Hall, G.M., Edn. 1st, Blackwell Publishing limited, New Jersey, USA, pp. 236-265.

Slizyte, R., Dauksas, E., Falch, E., Storro, I. and Rustad, T., 2005a. Yield and composition of different fractions obtained after enzymatic hydrolysis of cod (*Gadus morhua*) byproducts. *Process Biochem.*,40(34): 14151424.

Slizyte, R., Dauksas, E., Falch, E., Storro, I. and Rustad, T., 2005b. Characteristics of protein fractions generated from hydrolysed cod (*Gadus morhua*) by-products. *Process Biochem.*,40: 2021-2033.

Solina, M., Johnson, R.L. and Whitfield, F.B., 2007. Effects of soy protein isolate, acid-hydrolysed vegetable protein and glucose on the volatile components of extruded wheat starch. *Food Chem.*, 104(4): 1522-1538.

Song, R., Wei, R., Zhang, B. and Wang, D., 2011. Optimization of the antibacterial activity of half-fin anchovy (*Setipinna taty*) hydrolysates.*Food Bioprocess Technol.*, doi. 10.1007/s11947-010-0505-3.

Turpeinen, A.M., Kumpu, M., Ronnback, M., Seppo, L., Kautiainen, H., Jauhiainen, T. and Korpela, R., 2009.Antihypertensive and cholesterol-lowering effects of a spread containing bioactive peptides IPP and VPP and plant sterols.*J. Functional Foods*, 1(3): 260-265.

Vercruysse, L., Van Camp, J. and Smagghie, G., 2005. ACE inhibitory peptide derived enzymatic hydrolysates of animal muscle protein: a review. *J. Agric. Food Chem.*, 53(21): 8106-8115.

Wijesekara, I. and Kim, S., 2010. Angiotensin-I-Converting Enzyme (ACE) inhibitors from marine resources: prospects in the pharmaceutical industry. *Mar. Drugs*,8: 1080-1093.

Yoshikawa, M., Fujita, H., Matoba, N., Takenaka, Y., Yamamoto, T., Yamuchi, R., Tsuzuki, H. and Takabata, K., 2000. Bioactive peptides derived from food proteins preventing life-style related diseases. *Biofactors*, 12: 143-146.

Chapter 6

Marine Ornamental Fishes in Ratnagiri Coast, Maharashtra

☆ *N.D. Chogale, S.B. Satam, S.Y. Metar and H. Singh*

INTRODUCTION

Marine ornamental fishes are becoming the most popular attraction with colourful due to their adaptability to live in aquarium environment. The total trade of marine ornamental fishes has shown a steady increase over the past few years and varied between 120 to 420 cores. The marine ornamental fishes are abundant in the region which are rich in seaweeds, seagrasses, rocky bottom and corals in the seas. The variety of shapes, size colours and behaviors has exhibits by the marine ornamental fishes is amazing. These fishes are classified into more than 100 different facilities and most of them are bony fishes and few are cartilaginous. Total of 1471 species of ornamental fishes are graded globally.

Ratnagiri has a coast line about 182 km and very few survey has been conducted to assess the availability of the ornamental fish resources. However, collection of marine ornamental fishes revealed that there are 13 marine ornamental fishes belonging 12 families from rocky and mangroves areas of the Ratnagiri coast. The behaviors, colour patterns, appearance of collected marine ornamental fishes from the Ratnagiri coast are described in Table 6.1.

Distribution of Marine Ornamental Fishes

Heniochus acuminatus, Wimpal Fish or Banner Fish

Heniochus are among the delicate of all marine fishes. It has the longer snout, rounder shape and longer and more angular anal fin. Its body is compressed laterally,

Table 6.1: Collected Marine Ornamental Fishes from the Ratnagiri Coast

Sl.No.	*Family*	*Popular Name*	*Species*
1.	Chaetodontidae	Wimpal or Bannerfish	*Heniochus acuminatus*
2.	Chaetodontidae	Butterfly fish	*Chaetodon collare*
3.	Pomocanthidae	Bluering angle fish	*Pomacanthus annularis*
4.	Balistidae	Blue Trigger fish	*Odonus niger*
5.	Acanthuridae	Surgeon fish	*Acanthurus mata*
6.	Muraenidae	Laced moray eel	*Gymnothorax favagineus*
7.	Ambassidae	Glass fish	*Ambassis commersoni*
8.	Pomacentridae	Moon Tail Damsel	*Neopomacentrus filamentosus*
9.	Cichlidae	Pearl spot	*Etroplus suratensis*
10.	Scatophagidae	Spotted scat	*Scatophagus argus*
11.	Ephippidae	Teira batfish	*Platax teira*
12.	Monodactylidae	Silver moony	*Monodactylus argenteus*
13.	Diodontidae	Spotfin porcupine fish	*Diodon hystrix*

***Heniochus acuminatus*, Wimpal Fish or Banner Fish**

the first rays of its dorsal fin stretch in a long white filament. The background color of its body is white with two large black diagonal bands. Beyond the second black stripe, the dorsal and the caudal fins are yellow. The pectoral fins are also yellow. The head is white, the eyes are black and linked together by a black band. The snout, spotted with black, is a bit stretched with a small terminal protractile (it can be extend) mouth.

Chaetodon collare, Butterfly Fish

The red-tailed butterfly fish is also known as the Pakistani butterfly fish. The butterfly fish is brown to black, with lighter scales giving it a spotted appearance. It has a prominent, vertical white streak behind the eyes, a dark stripe over the eyes, and another, smaller white stripe in front of the eyes. The base of the tail is bright red, followed by a black stripe. The tip of the tail is diffuse white.

***Chaetodon collare*, Butterfly Fish**

Pomacanthus annularis, Bluering Angle Fish

Blue ring angelfish can undergo changes in coloring in its development from youth to maturity. The young of the blue ring angelfish are at first a dark blue, almost black, with broad turquoise and white vertical stripes. Adults will exhibit a violet coloring with semicircular marks on the middle of the body.

***Pomacanthus annularis*, Bluering Angle Fish**

Odonus niger, Blue Trigger Fish

The red toothed triggerfish is a dark blue bodied fish, ranging up to 50 cm in length. The fins are all blue-green in color, having yellow and light blue trim. It has a lyre shaped caudal fin with a yellow bar between the lobes. Like all triggerfish red toothed triggers have a retractable dorsal spine. Their pectoral fins are quite small; as a result they steer mostly with their dorsal and anal fins, which makes them very maneuverable, and they also use these fins to move with an exotic type of propulsion reminiscent of a propeller. It is one of the most singular swimming styles in the ocean.

Acanthurus mata, Surgeon Fish

The body has an oval shape and is compressed laterally. The caudal fin has a crescent shape. The mouth is small and pointed. Its body is streaked with horizontal bluish lines on a brown background color although over time it is able to change colour to become grey-blue overall. A longitudinal yellow stripe runs across the eye and splits in two lines extending anterior the eye. The superior lip is also yellow. The dorsal and anal fin are bluish with a yellow reflection, the base of latters is underlined by a fine black line. The slit of erectile scalpelis darker. Its erectile spine, sharp as surgeon's scalpel, located at the base of the tail is a dread defensive weapon.

***Odonus niger*, Blue Trigger Fish**

***Acanthurus mata*, Surgeon Fish**

Gymnothorax favagineus, Laced Moray Eel

Gymnothorax favagineus is serpentine in shape body has a white to yellowish background color dotted with numerous black spots which latter vary in size and shape depending on the individual and on the environment in which the animals

***Gymnothorax favagineus*, Laced Moray Eel**

live. Therefore, morays living on a reef with clear water will have less black spots than those of a turbid environment. It is from this characteristic color pattern that ensue its vernacular names.

Ambassis commersoni, Glass Fish

These are elongated and laterally compressed fishes with glassy or semi transparent body. They are move in shoals and are good to be kept in aquaria. This fish inhabits in the estuarine and coastal waters along the coast of India.

***Ambassis commersoni*, Glass Fish**

Neopomacentrus filamentosus, Moon Tail Damselfish

Adults are with long extended lobes on the dorsal and caudal fins. Body colour is plain grey with greenish shine over back. Twin yellow spots on operculum, just above pectoral fin base in juveniles. Fins margin blue narrow. Anal fin is whitish.

***Neopomacentrus filamentosus*, Moon Tail Damselfish**

Etroplus suratensis, Pearl Spot

These are deep bodied light green coloured fishes with 6 to 8 vertical black bands, and having white pearly spots all over the body. They can grow up to 40 cm in natural environment; but small ones are preferred in aquaria.

***Etroplus suratensis*, Pearl Spot**

Scatophagus argus, Spotted Scat

The body is strongly compressed. The dorsal head profile is steep, with a rounded snout. The body is greenish-brown to silvery with many brown to red-brown spots. Spines and rays of the dorsal fin are separated by a deep notch. Small ctenoid scales cover the body. Juveniles are a greenish-brown with either a few large, dark, rounded blotches, or five or six dark, vertical bars.

***Scatophagus argus*, Spotted Scat**

Platax teira, Teira Batfish

Platax teira has a dark blotch under the pectoral fin, with another long dark mark above the base of the anal fin. Looked at from the side, it has a roughly circular body with a low hump on the nape. This fish is usually silver, grey or brownish. It has a blackish band through the eye and another band with the pectoral fin.

Monodactylus argenteus, Silver Moony

The silver moony has a silver body in a triangular shape. The big eyes are crossed by a dark ray. Its fins are silver with yellow reflection. Both, yellow and dark colouring, are fading with age. It is also confused, though rarely, with *Monodactylus sebae*, but easily distinguished by its other black ray that crosses where its tail starts, darker color and less yellow coloration. It is commonly called finger fish because its genus Monodactylus means "one finger". It is also sometime called "Sea Angel" in the pet shops.

***Platax teira*, Teira Batfish**

Diodon hystrix, Spotfin Porcupine Fish

Body is elongated with a spherical head with big round protruding eyes, a large mouth rarely closed. The pectoral fins are large, the pelvic fins are absent, the anal and dorsal fins are close to the caudal peduncle. The latter move simultaneously during swimming. The skin is smooth and firm, the scales are modified into spines. The body coloration is beige to sandy-yellow marbled with dark blotches and dotted with numerous small black spots.

In case of danger, the porcupine fish can inflate itself by swallowing water to deter the potential predator with its larger volume and it can raise its spines. The porcupine fish concentrates a poison, called tetrodotoxin, in certain parts of its body such as the liver, skin, gonads and the viscera. Tetrodotoxin is a powerful

***Monodactylus argenteus*, Silver Moony**

***Diodon hystrix*, Spotfin Porcupine Fish**

neurotoxin. This defensive system constitutes an additional device to dissuade the potential predators.

REFERENCES

Chogale, N. D. and Bhatkar V. R. 2006. Some Important marine ornamental fishes of Ratnagiri coast. *Fishing Chimes*. 25(12)34-36.

MPEDA, 1994. Hand book on Aqua farming: Ornamental fishes.

Murugan, M., Sahu M. K., Srinivasan, M., Devi, K., Khan S. A. and Kannan L., 2010. Marine ornamental fishes in the little Andaman island. In: Advances in Aquatic Ecology (Vol 3), V.B.Sakhare (ed.), Daya Publishing House, Delhi, 12-18.

Pradeep, B. 2004. Neglected ornamental fish resources of Kerala. *Fishing Chimes* vol. 24, No. 3, 59-60.

Riseley, R. A. 1971. Fishes. In: Tropical Marine Aquaria The Natural System. Riseley, R. A. (Ed.). George Allen and Unwin Ltd., London, 83-109.

Satam S. B., Chogale N. D., Chavan B. R., Metar S. Y., Sadavarte V. R., Sawant A. N. and Singh H. 2014. Availability of ornamental fishes in mangroves of Ratnagiri. In: National Seminar on Mangroves of Konkan coast held at Gogate and Jogalekar College, Ratnagiri on 2nd and 3rd February 2015.

Satam S. B., Sadawarte V. R., Sawant A. N., Chogale N. D., Metar S. Y. and Chavan B. R. 2015. Ornamental fish biodiversity in Mirya creek, Ratnagiri. *Ecology and Fisheries*. 8(1): 32-36.

www.fishbase.org

Chapter 7

Discus: The King of Aquarium Fish

☆ *S. Nair, Sachin Satam, P.E. Shingare, A.U. Pagarkar and N.D. Chogale*

The Discus Fish (*Symphysodon aquefasciatus*) is one of the most celebrated ornamental fishes today. The discus is essentially a "round disc-shaped" fish whose natural habitat is the Amazon basin of South America. The Discus is called "king of aquarium fish" Why? What is so special about Discus, which makes Discus so special? Discus is the only species of fish that is born with the ability to excrete food in the form of mucus from whole body to feed their young, quite similar to our human breast feeding. In human only mother can produce milk where as in Discus both parents can produce this food for their babies. This makes Discus the only living thing in this world gifted with this unique capability.

In olden days only wild strains were available but today many more and more discus strain had developed, many discus competitions are taking place globally. Today discus is not just an aquaculture commodity. It has become multi- million dollar fashion industry with many designers (breeders) marketers, shows and consumers who are anxiously waiting for next trendy design.

Tank Setup for Discus

The aquarium setup for discus fish is generally one to two types, bare-bottom or planted. Whichever set-up you choose, one thing is for sure, it needs to be large. Actually bare bottom tanks are mostly preferred which is easier to clean and minimize the contamination. Since Discus is schooling fish, they are best kept in groups of at least 3-4.

Discus fish need water that is slightly warmer than many tropical fish, the best temp range for discus is 28-30°C. If the temp is below the required temperature then a proper heater and thermometer are a must for a discus aquarium.

Filtration is the next area to consider. Generally biological filters are preferred over chemical or mechanical filters. A good biological filter allows lots of good nitrifying bacteria to colonize it and remove toxins and impurities from the water.

If going with a planted tank its best to try a best to decorate the tank as natural as possible to stimulated normal behaviour and reduce stress in your discus. An aquarium with large plants, some rocks, roots and a few floating plants is the ideal setup for the discus. It provides hiding places filters the light from above and contributes to a healthy water condition. Any substrate used in the bottom of a discus aquarium should be inert so that it does not affect the pH level and hardness of the water. Avoid high intensity lighting as the discus fish will not like. A well planted aquarium properly setup is a thing of beauty and you will spend countless hours just watching your discus as they exhibit their natural behavior in their beautiful aquarium.

Water Required for Discus

To keep your discus healthy your focus on their water should be in two areas, cleanliness and balance. Discus fish require regular water changes. Water changing simply removes old, stale water from the discus fish aquarium and replaces it with freshwater. Water changes remove impurities from the discus aquarium including the end products of nitrification (the biological activity of bacteria consuming the waste products in the water). Discus fish like clean water in their aquarium. More importantly, discus fish require clean water in their aquarium in order to remain healthy. A proper biological filter should also be installed. Water changes in Discus Tank is directly related to the density of fish in the tank, type of feeding (Beef Heart, live feed, Frozen *etc.*), no. of feeding per day, filtration system and incoming water parameter.

Keeping your discus fish water properly balanced is the next step for proper water conditions. Obviously a good chlorine and chloramines neutralizer is necessary to remove these chemicals which are deadly to all fish. But keeping your water at the right ph level and hardness level are also important to the health of your discus fish, especially if you want to breed discus. As a general rule discus like softer water, but this is less true today than in the past, as many generations of breeding discus in have adapted them to water conditions that a quite different from the native water in the Amazon.

Nutrition

Quite a bit of knowledge has been gained about the discus fish diet from studying the stomachs of wild discus. In the wild discus eat small shrimp, insects and insect larvae, small fish and fair amounts of plant material. If you ever watched a discus eat, you probably noticed that they spit their food out and then recapture it and swallow it. The reason for this is their teeth, or lack of teeth. Discus fish do not have teeth, but rather grinders in the jaw. The scientific name "symphysodon" means jaw grinders. So when the discus fish is eating, food is taken into the mouth, ground up and spit out, then picked up again for regrinding before swallowing. During this process, many food particles fall to the bottom of the discus fish aquarium. Discus fish will expel water at foods resting on the aquarium bottom, and then pick them up as they float.

Many articles published by various discus breeders tell us that discus is very flexible in their dietary preferences regarding their discus fish diet. Some discus fish breeders feed single source foods such as brine shrimp, beef heart or bloodworms. Other discus fish breeders feed live foods. The average discus fish breeder uses a combination of frozen foods and dry foods.

Feeding live foods such as tubifex worms, daphnia, bloodworms, mosquito larvae accelerate the growth in Discus, but they carries an element of risk and are not safe food for discus. These live foods come from polluted sources and they contain a host of parasites, bacteria and viruses most of which are very difficult to eliminate with the best of medicines or chemicals.

When such unsafe live food are feed to the Discus, the pathogens are passed on to the parents and then to their frys and finally the fish dies. This is very disappointing situation to the hobbyist and breeders, due to which some times the hobbyist shy away from the hobby. In any event, no matter how you feed your discus, you should make sure they get Proteins, Fats, Carbohydrates, Vitamins and Minerals in correct proportion in their diet.

Tips on Sexing Discus Fish

Here are ways to determine the sex of your discus:

1. **The dorsal and anal fins:**Take a good look at the dorsal and anal fins of your discus. Are they rounded or are they pointed. Males tend to have pointed dorsal and anal fins or sometimes have extended growth on the end rear of the dorsal fin. Females tend to have rounds rear fins.
2. **The colour and pattern of your discus:** Please look at the colour and pattern of your discus fish, compare them to one another in the tank. Some fish will hopefully have more intense colour and some discus will have more pattern. Male discus fish tend to have less intense colour but have more pattern while the female tends to be more colorful but with lesser pattern.
3. **Size of the discus:** Compare the size of your discus. Male discus tend to be bigger than the females but the size can also depend on whether the fish was stunted or are just small in genetic make up.First of all, know that these are only guides and they are not always 100 per cent correct as in some circumstances females have male characteristics and males have female ones.

How to get a Breeding Pair of Discus Fish

Breeding Discus is one of the most exciting things you can do as a discus owner. Nothing is more thrilling to a discus enthusiast than watching two of his discus pairs off, go through their mating ritual, spawn and then hatch a load of baby discus.

Figure 7.1: Hatchery Set-up for Discus Breeding.

Figure 7.2: Red Passion Discus Pair with Eggs on Breeding Cone.

Figure 7.3: Discus Pair with Babies on Breeding Cone.

Most discus keepers will one day want to breed there fish. Some will just want one pair and some will want there very own fish house stocked full of discus fish.

Everyone starts with just one pair, I know I did. I soon wanted more and then more. As you go along you keep adding to your collection. The good thing about breeding discus is that you can cover all of your expenses by selling your young one.

Now I know some of you are just getting into discus fish and might like to know how is the best way to go about breeding discus and getting a pair.

Figure 7.4: Parental Care.

Figure 7.5: Leopard Discus.

Figure 7.6: Pigeon Discus.

(*All Figure Courtesy*: Mr. Alhad Phadnis, Pune)

There are three options available if you want to acquire a pair of discus, here there are:

1. Firstly, the most common way to do it is to buy six to eight young discus at around 3 inch and grow them up. By doing this you should be able to get one or two pairs of discus.
2. Secondly you can buy a proven pair of discus from an importer. This is risky as you need proof and they will be very expensive.
3. The last way to get a breeding pair of discus is to buy two or three young adult discus at around 5in size. At this size there may be some signs of what sex the fish is and if you get three your chances of getting a pair can be good.

Now there are positives and negatives about all of these methods. Not one method is better than the next but it depends on the individual. For a beginner I would recommend starting which young less expensive fish as then you have room for mistakes.

Keeping Discus Fish Disease Free

One of things you must do is keep your discus fish healthy and disease free. One of the most common assumptions with discus is that they easily get disease and die. This is not true. They will only get disease if you mistreat them.

The root of the problem is stress; if you cause discus stress there immune system will lower and make them venerable to disease. It is important to keep stress to a minimum so below a checklist of things to do is given:

- ☆ Make sure all water is free from chlorine and heavy metals. It is important that you invest in either an RO unit, water purifier or use a good quality tap water condition.
- ☆ Give your discus a good varied diet.
- ☆ Don't place your discus aquarium near a door and all the movement will cause them stress.
- ☆ If you are keeping young discus, keep in groups of no less than six. Discus is a shoaling fish.
- ☆ Make sure there are no large changes in pH, hardness or temperature when changing water.

Above are just a few things that you can do to make sure your discus fish live a happy life and remain healthy. Remember discus is not hard to keep, they just require regular water changes and a good varied diet like most other fish.

Chapter 8

Ornamental Fish Biodiversity in Mirya Creek of Ratnagiri, Maharashtra

☆ *S.B. Satam, V.R. Sadawarte, A.N. Sawant, N.D. Chogale, S.Y. Metar and B.R. Chavan*

ABSTRACT

The tropical marine ornamental fishes play an important role in the ornamental fish trade and they are now a popular subject of research. Conservation of ornamental fish biodiversity is most important issue for improving livelihood of low income coastal communities. The survey of ornamental fishes was performed at Mirya creek of Ratnagiri for study of biodiversity during Septemeber 2013 to April 2014. It lies between approximately 17° 02′ 20″ N lat. and 73° 27′ 13″ E long. A total of 30 species under 5 orders, belonging to 21 families and 27 generas were collected from the study area. The aim of the present investigation is to enumerate the ornamental fish biodiversity in Mirya creek of Ratnagiri.

Keywords: *Ornamental fish, Biodiversity, Mirya creek, Ratnagiri, Maharashtra.*

INTRODUCTION

Biodiversity plays a key role in the environment through biogeochemical process that sustains the biosphere. Biological biodiversity refers to the living organisms from all sources, including, inter alia, terrestrial, marine and other aquatic ecosystems, (Ravi, 2013). The ornamental fishes can be defined on the basis of their attractive colourations, peaceful nature, tiny sizes, its suitability for keeping in captivity and adaptability for living in confined spaces (Dey, 1997). The estuaries and coastal waters are the richest areas of ornamental fish resources. One of the threats to this natural resource is pollution due to domestic sewage, industrialization and other anthropogenic activities (Sahib, 2013).

Materials and Methods

The present work was carried out during Septemeber 2013 to April 2014. The fishes were collected at monthly intervals from Mirya creek of Ratnagiri with the help of hook and lines, cast nets and also from local fisherman. The fishes were immediately noted down for their colour and other morphological features. The fishes were identified up to the species level using standard keys of Talwar and Jhingran (1991) and IUCN (2009).

Enumeration

The ornamental fishes with their order, family, scientific name, common name, local name (marathi), distribution and fishery information are given below:

1. *Acanthurus mata,* Cuvier, 1829, (Acanthuridae), Elongate surgeonfish, Marathi Name:Mata Tang. This fish is widespread all over the Indo-Pacific region. It inhabits around coral reefs in depth range from 5 to 45m. Maximum size of fish recorded 50 cm in length.
2. *Apogonichthyoides pseudotaeniatus,* Gon, 1986, (Apogonidae), Double bar cardinal fish, Marathi Name: Aamta. This fish distributed in Indo-West Pacific and reaches maximum size of 14 cms.
3. *Apogon endekataenia,* Bleeker, 1852, (Apogonidae), Candystripe cardinal fish, Marathi Name:Aamta. Distribution: Occurs along cliffs and slighty deep water. This species is nocturnal.
4. *Ambassis commersoni,* Cuvier, 1828, (Ambassidae), Glass fish, Marathi Name: Kachaki. This fish inhabits in the estuarien and coastal waters along the coast of India. Considered as a food fish and also used as bait for catching other commercially important fishes. This species grows upto 16 cm size.
5. *Etroplus suratensis,* Bloch, 1790, (Cichilidae), Pearl spot, Green Chromide, Marathi Name: Kondawa. Inhabits in brackishwater and mouth of rivers. This species grows upto 40 cm size. A delicious fish, extensively cultivated in ponds and tanks. They thrives well as well as breed easily, where growth of aquatic vegetation found.
6. *Chaetodon collare,* Bloch, 1787, (Chaetodontidae), Red tailed butterfly fish, Marathi Name: Chand. It can be found in reefs of the Indo-west Pacific region. It usually swims at depths of between 3 and 15 m. It can grow upto 18 cm in size. This fish is brown to black, with lighter scales giving it a spotted appearance.
7. *Heniochus acuminatus,* Linnaeus, 1758, (Chaetodontidae), Longfin banner fish, Marathi Name: Kombada. It is wide spread in tropical Indo-West Pacific. It attains maximum size of 45 cm in length.
8. *Platax orbicularis,* Forsskal, 1775, (Ephippidae), Orbicular batfish, Marathi Name: Kawala. This species is inhabitant in brackish as well as marine

waters at depth ranging from 5 to 30 m in the tropical waters of the Indian and Pacific Oceans. Juvenile fishes lives in small groups, among mangroves. The body is almost disc-shaped, and very thin. Its tail is fan shaped and is taller than its length which is about 20 per cent of the body length. This fishes can grow to up to 50 cm in length.

9. *Platax teira*, Forsskal, 1775, (Ephippidae), Longfin batfish, Marathi Name: Kawala. This species is distributed in Indo-West Pacific region. This fishes found in shallow coastal habitats to deeper offshore. This fish is usually silver, grey or brownish in colour. It has a blackish band through the eye and another band with the pectoral fin. This fishes can grow to up to 24 cm in length.
10. *Gerrus filamentosus*, Cuvier, 1829, (Gerreidae), Whipfin silverbiddy, Marathi Name: Shetuk, Charbat. This species is wide spread in Indo-West Pacific, found in shallow waters, usually in shoals. This fishes are caught in trawls, boat-seines and traps. This fish average size is 15 cms and can grows upto 25 cms.
11. *Halichoeres biocellatus*, Schultz, 1960, (Labridae), Red-lined Wrass, Parrot fish, Marathi Name: Popat. The IUCN Red List of threatened species lists this species to be of least concern. Although populations may be declining slowly in Southeast Asia where fishes are collected for aquarium trade. This fishes can grow to up to 12 cm in length.
12. *Monoactylus argenteus*, Linnaeus, 1758, (Monodactylidae), Monoangel, Silver Moony, Marathi Name: Chandlak. This species common in the coastal waters, and wide spread in Tropical Indo-West Pacific. This fish can live in captivity for a number of years, and attains size upto 25 cm in length. They are caught with beach seines and cast nets.
13. *Chrysiptera unimaculata*, Cuvier, 1830, (Pomacentridae), One Spot Damsel, Marathi Name: amsel. This species is wide spread in Tropical Indo-West Pacific, common in the coastal waters. This fishes can grow to up to 9 cm in length.
14. *Chrysiptera parasema*, Swainson, 1839, (Pomacentridae),Yellowtail damsel, Marathi Name: Damsel. This damselfish is native to the western Pacific Ocean. It is bright blue with a yellow tail. This fishes can grow to up to 8 cm in length.
15. *Neopomacenprus filamentosus*, Macleay, 1882, (Pomacentridae), Moontail damsel, Brown demoiselle, Marathi Name: Damsel. This species is wide spread in Tropical Indo-West Pacific, common in the coastal waters. This fishes can grow to up to 9 cm in length.
16. *Abudefduf saxatilis*, Linnaeus, 1758, (Pomacentridae), Sergant major, Marathi Name: Kondwa. This species is wide in tropical and subtropical locations found at depths of 0 to 20 metres. This fishes can grow to up to 22.9 cm in length.
17. *Pomacanthus annularis*, Bloch, 1787, (Pomacanthidae), Bluering angel fish, Marathi Name: Nila Angel. This species is wide spread in Tropical Indo-

West Pacific. Juvenile bluering angelfish are at first a dark blue with broad turquoise and white vertical stripes, whereas adult exhibit a violet colors with semicircular marks on the middle of body. This fishes can grow to upto 30 cm.

18. *Siganus canniculatus*, Park, 1797, (Siganidae), Whitespotted spinefoot, Marathi Name: Mutri. Occurs in in Tropical Indo-West Pacific in schools in coastal waters at depth of 40 cm. Caught in shore seines, bottom trawls, set nets and traps. This fishes can grow to upto 23 cm.
19. *Siganus vermiculatus*, Valenciennes, 1835, (Siganidae), Vermiculated spinefoot, Marathi Name: Tawus. Occurs in Indo-west Pacific in schools in coastal waters at depth of 40 cm. Caught in shore seines, bottom trawls, set nets and traps. This fishes can grow to up to 45 cm.
20. *Scatophagus argus*, Bloch, 1788, (Scatophagidae), Scat, Marathi Name: Wada. This species inhabits estuaries and mangroves areas and common in Indian backwaters. Young ones are beautifully coloured and used as handsome aquarium fishes. This fishes are caught mainly with the traps and the gillnets. This fishes can grow to upto 30 cm.
21. *Pelates quadrilineatus*, Bloch, 1790, (Teraponidae), Fourlined terapon, Marathi Name: Hajam, Garangat. This species is wide spread in Indo-west Pacific, inhabits in marine and brackishwaters. Common near the mouth and brackishwaters. This fishes can grow to upto 20 cm.
22. *Terapon jarbua*, Forsskal, 1775, (Teraponidae), Jarbua Terapon, Marathi Name: Hajam, Garangat. This species is wide spread in Indo-west Pacific, founds in shore waters and moves considerable distance upstream into freshwater. This is the most common species, good as aquarium fish, caught all along the coast of India. This fishes can grow to up to 30 cm.
23. *Gymnothorax favagineus*, Bloch and J. G. Schneider, 1801, (Muraenidae), Laced moray eel, Marathi Name: Wam/Rechan. This species is widespread in the Indo-West Pacific. This species can reach a maximum length upto 300 cm, but specimens usually encountered are much smaller. This fish are with serpentine in body shape, has a white to yellowish background color, dotted with numerous black spots.
24. *Pempheris vanicolensis*, Cuvier, 1831, (Pempheridae), Vanikoro sweeper, Marathi Name: Bamata. This species distributed in the Central Indo-Pacific, in rocky shores and coral reefs. This fishes can grow to up to 15 cm in length.
25. *Oryzias melastigma*, McClelland, 1839, (*Oryziadae*), Indian Medaka, Rice Fish, Marathi Name: Piku. Inhabit primarily in brackish water, freshwater ponds, lakes and rivers. They mostly occur throughout South East Asia. Distributed in West Bengal, Kerala, Tamil Nadu and some parts of Maharashtra as well as Karnataka. They are generally found in gutters, paddy fields, foul water, brackish water. It is perennial breeder, which grows maximum up to 4.0 cm in length and attains maturity within two to three months at about 1.6 cm length.

26. *Epinephenus dicanthus*, Valenciennes, 1828, (Serranidae), Six barred reef cod, Marathi Name: Hekaru, Gobra. This species is wide spread in the Central Tropical Indo-Pacific. This fishes can grow up to 46 cm in length.
27. *Odonus niger*, Ruppell, 1840, (Balistidae), Redtoothed triggerfish, Marathi Name: Nili Kend. This species found in the Tropical Indo-Pacific and can grow up to 20-30 cm in size.
28. *Diodon hystrix*, Linnaeus, 1758, (Diodontidae), Spotfin porcupine fish, Marathi Name: Kend, Fugu. This species is found in all the tropical and subtropical waters of the world. Juveniles are pelagic up to the time that they are about 20 cm in length. Adults favour lagoons, top reefs and rocky reefs from one to 50 m depth, sheltering under ledges or in caves during the day. In case of danger, the porcupinefish can inflate itself by swallowing water to deter the potential predator with its larger volume and it can raise its spines. These fishes concentrates a powerful neurotoxin named as 'tetrodotoxin', in certain parts of its body such as the liver, skin, gonads and the viscera. This fishes can grow up to 40 cm in length.
29. *Hippocampus kuda*, Bleeker, 1852, (Sygnathidae), Yellow Seahorse, Marathi Name: Sagari Ghoda. This species found in a shallow coastal waters of the Indo-Pacific, including coral reefs, muddy slopes, and shallow estuaries. This fishes are poor in swimming, so they spend most of their time anchoring to coral reefs and branches with their tails. Hence they need anchor points in aquaria. Also they like a quiet tank, without large, belligerent fish, and a slow-moving current. This fish is the one of most traded species in many Southeast Asian countries for traditional chinese medicine and its conservation status is currently listed by the IUCN as 'vulnerable'. As per the CITES, all species of seahorse have been placed under Appendix II.
30. *Pterois volitons*, Linnaeus, 1758, (Scorpaenidae), Lion fish, Marathi Name: Kombada. This species is widespread in the Indo-Pacific, a predator and aggressively preys on small fish and invertebrates. This fishes found around the seaward edge of reefs and coral, in lagoons, and on rocky surfaces to 50 m depth. They are well known for their ornate beauty and venomous spines.It can grow up to 45 cm in length.

Results and Discussion

A total of 30 species under 5 orders, belonging to 21 families and 27 genera were collected from the Mirya creek of Ratnagiri during the present study period. Jhingran, (1991) reported that the major factors determining the distribution of fishes in the estuaries are temperature and salinity. Marine fishes dominated the Mirya creek which could be justified by the fact that the creek is connected to the sea and the salinity is high throughout the year, however Divakaran *et al.* (1982) were reported same in Thevally kadavu lake region. It has been observed in other estuaries of India that marine fishes migrate into the estuary where the hydrobiological conditions are favourable (Sinha and Nandan, 1996).

It was found that species such as *Etroplus suratensis, Ambassis commersoni, Chaetodon collare, Monodactylus argenteus, Chrysiptera unimaculata, Chrysiptera parasema, Abudefduf saxatilis, Neoponacenprus filamentosus, Oryzias melastigma, Odonus niger, Terapon jarbua, Scatophagus argus, Gerras filamentous* occurred uniformely throughout the study period. The information collected indicates that around 30 species from Mirya creek are having ornamental characteristics, can be utilised for aquarium for improving livelihood of low income coastal communities. But also there is a need to conserve the natural resources of the creek from human disposal of untreated waste as well as water.

Acknowledgements

The authors are thankful to the Senior Scientific Officer and Research Officer of Marine Biological Research Station, Ratnagiri for the facilities provided.

REFERENCES

CITES, 2014, In: Appendices I, II and III valid from 14th Sept. 2014, www.cites.org, Geneva, Switzerland.

Dey, V.K., 1997. In: Handbook of Aquafarming, Manual, MPEDA, Kochi, 76 pp.

Diwakaran, O.M., Arunchalam, Nair N.B. and Balasubramania, N.K. 1982. Seasonal variation of zooplankton of the Ashtamudi estuary, South West Coast of India. *Mahasagar. Bulletin of the National Institute of Oceanography*, 1(1): 43-50.

IUCN, 2009. IUCN Red List of Threatened Species, Version 2009. (www.iucnredlist.org).

Jhingran, V.G., 1991. Fish and Fisheries of India, Hindustan Publications, Delhi, 727 pp.

MPEDA, 2008, In: An illustrated guide on commercial fishes and shellfishes of India, MPEDA, Kochi, 187 pp.

Ravi, V., 2013, Marine biodiversity of India, In: Emerging Trends in Fisheries and Aquaculture (Eds.V.B.Sakhare and B.Vasanthkumar), Daya Publishing House,New Delhi, pp. 1-25.

Sahib, S. S., 2013. Fish diversity in the Thevally Kadavu Lake, Kollam. In: Emerging Trends in Fisheries and Aquaculture (Eds.V.B.Sakhare and B.Vasanthkumar), Daya Publishing House, New Delhi, pp. 95-105.

Sinha and Nandan, D.S., 1996. A preliminary survey of the fishery resource of the Asthamundi estuarine system. In: Ecology of Indian Estuaries, *Fish. Techn.*11 (20): 75-83.

Talwar, P.K. and Jhingran, A.G.1991. In: Inland Fishes of India and Adjacent Countries, Vols.1 and 2. Oxford and I.B.H. Publication., New Delhi.

Chapter 9

Introduction of some Colour Enhancers in Ornamental Fishes

☆ *M.M. Girkar and S.N. Kunjir*

INTRODUCTION

Ornamental fishes are characterized by a wide diversity of colours and colour patterns and success in the ornamental fish trade is very much dependent on the vibrant colour of the fish. Colour is one of the major factors, which determines the price of aquarium fish in the world market. Fish are coloured in nature often show faded colouration under intensive culture conditions. Fish like other animals do not synthesize carotenoid and depend on dietary carotenoid content for the colouration. Hence, a direct relationship between carotenoids and pigmentation exists in them.

Carotenoids and carotenoid-protein complexes are the main source of fish skin and muscle pigmentation; therefore, to increase the skin and flesh colour in captivity, fish must obtain an optimum level of carotenoids in their diet. Fish skin colour is mainly attributed to the presence of chromatophores that contain pigments including melanin, pteridines, purines and carotenoids.

Carotenoids in Fishes

Carotenoids commonly occurring in fishes with their colours are depicted in Table 9.1.

Purpose of Colour

The main purpose of colour in fish involved for following purpose:

1. Communication
2. Identification

Table 9.1: Common Carotenoids Occurring in Fishes with their Colours

Carotenoids	*Pigment Colour*	*Fish Species*
Tunaxanthin	Yellow	Scombrina, Carangina and Percina but abundantly in yellow colouerd fishes like yellow tail
Lutein	Greenish-yellow	Commonly in freshwater fishes but widely found in marine fishes
Beta- carotene	Orange	—
Alpha, Beta-doradexanthins	Yellow	—
Zeaxanthin	Yellow-orange	Anemone fish
Canthaxanthin	Orange red	—
Astaxanthin	Red	Pink colouration in salmon, Serpae tetra
Eichinenone	Red	—
Taraxanthin	Yellow	—

3. Camouflage
4. Defense
5. Mimicry

Plants Based Natural Carotenoid Pigments

The major sources of plants based carotenoid pigments are *Chlorella vulgaris, Haematococcus plurialis, Spirrlina* sp. shrimp head meal, marigold, paprika, carrot, pumpkin, capsicum, wheat corn, alfalfa, krill extract, cray fish and red yeast (*Phaffia rhodozyma*).

Animal Based Natural Carotenoid Pigments

The major sources of animal based carotenoid pigments are antarctic krill, crayfish meal, shrimp meal, crab meal and some microorganisms.

Colour Enhancement Methods in Ornamental Fishes

The colour of ornamental fish may be enhanced by following methods:

1. Dyeing
2. Painting
3. Feeds with hormone
4. Feeds with pigments

Dyeing

Dyeing colorless fish has recently become popular. The fish are immersed in water containing dye and the immersion and handling may lead to disease problem.

Painting

The practice of painting essentially colourless fish has become widespread. The neon coloured paint is non-toxic but the handling and painting, coupled with shipping stress often invites disease problems like ich and fungal infection. The paint is shed in times and the fish returns to being colourless.

Feeds with Hormone

Hormones may be used to enhance fish colouration by causing a false early maturity. Testosterone supplied in the diet likely allows a premature storage and expression of pigments in the chromatophores. Fish treated with hormones often become all male, sterile and require a continuous dietary supply of hormones to maintain coloration.

Feeds with Pigments

Carotinoids are the group of over 600 natural lipid soluble pigments that are primarily produced with in phytoplankton, algae and plants. These pigments are responsible for variety of colours in nature.

Top Colour Enhancing Compounds

1. **Spirulina:** This has an above average carotenoid content that is easily assimilated and found in simple algae cells.
2. **Paprika:** Red pepper meal is more potent than spirulina and contains red pigments ready to be absorbed and assimilated immediately.
3. **Marigold flower meal:** Probably one of the most potent natural sources of Carotenoids available. The pigments require manipulation by Koi carps to convert them to red colour.
4. **Artificial colour enhancers:** Astaxanthin and Canthaxanthin are guaranted, potent source of colour in the form that will exhibited immediately in the skin.
5. **Red yeast:** This is easily digested and is a recognized source of carotene and astaxanthin.

REFERENCES

Ahilan B. 2014.Colour enhancement in ornamental fish. Published in the International Seminar on Ornamental fish breeding, farming and trade, 26-27 January 2014, Cochin, India, organized in connection with India International Aqua Show 2014 held at Cochin.pp.67-73.

Choubert G.1979. Tentative utilization of *Spirulina* algae as a source of carotenoid pigmentation for rainbow trout, *Aquaculture*, 18: 135-143.

Gupta, S.K. A.K Jha and G Venkateshwarlu.2007. Use of natural Carotenoids for pigmentation in fishes. *Natural product Radiance*. 6(1): 46-49.

Minerva García-Chavarría and Maurilio Lara-Flores. The use of carotenoid in aquaculture. *Research Journal of Fisheries and Hydrobiology*, 8(2): 38-49, 2013.

Nakano T, Kanamuri T, Sato M and Takeuchi M. 1999.Effect of astaxanthin rich red yeast (*Phaffia rhodozyma*) on oxidative stress in rainbow trout. *Biochem. Biophys. Acta*.1426: 119-125.

Selvakumar Dharmaraj and Kandasamy Dhevendaran. 2011.Application of Microbial Carotenoids as a Source of Colouration and Growth of Ornamental Fish *Xiphophorus helleri*. *World Journal of Fish and Marine Sciences*, 3 (2): 137-144.

Chapter 10

Performance of Freshwater Exotic Fishes in India: A Review

☆ *V.B. Sakhare*

The first attempt to document fish introductions on a global scale was undertaken by FAO under the stewardship of Dr.Robin Welcome. There are 1354 introductions of 237 species into 140 countries. Out of 237 species recorded from 140 countries, it is interesting to note that only 9 species have gained entry to more than 30 countries. Among 9 species, it is the common carp, black tilapia and rainbow trout that have become pan-global in distribution, with each species being introduced in more than 50 host countries.

Shafland and Lewis (1984) defined an exotic as an organism whose entire range is outside the country to which it is introduced. Exotic species of fishes were introduced in many parts of the world for:1) improving local fishery potential and for broadening species diversity in aquaculture programmes,2) sport fishing,3) for aquarium keeping, and 4) controlling of unwanted organisms. Further, there are accidental and/or unauthorized introductions.

There is a long history of introduction of exotic fishes in freshwaters. For researchers, managers, and policy makers interested in conserving freshwater diversity, understanding the magnitude and array of potential impacts of exotic fish species is of utmost importance. The freshwater fish diversity is depleting alarmingly due to the introduction of exotic species and other anthropogenic activities. The indiscriminate transfer of exotic fishes has brought about a wide array of problems including extirpation of indigenous species. The exotics compete with the indigenous species for food, habitat and may even prey upon them, introduce new parasites and diseases, result in the production of hybrids and cause genetic 'erosion' of indigenous species and degradation of the physico-chemical nature of aquatic ecosystems. The potential risks not only affect the biodiversity, but also the socio-economic aspects of the human community that depend on aquatic ecosystems for their sustenance (Bijukumar, 2000).

Majority of species have been introduced into only a few countries. Eighty-six percent of the species introduced have been recorded from 10 countries or less.29 countries have only 1 species introduced, 6 countries have had only 2 species introduced and so on. Continental United States leads the list with a spectacular 70 species introductions followed by island state of Hawaii (44 introductions), Columbia (40), Mexico (33) and others (Ambekar *et al.*, 2008).

The ecological impact of exotic introductions resulted in the disappearance of native species in many countries. Out of 160 species of freshwater fishes introduced in 120 countries, hardly 10 per cent were found satisfactory. During 1950s, Lake Victoria, the world's largest tropical lake was the home of more than 300 species of endemic fishes. Due to introduction of an exotic fish, Nile Perch (*Lates niloticus*) which is a predator and voracious feeder, the species composition of the lake changed drastically with a loss of 99 per cent of the species diversity which came down to less than 1 per cent of the total fish catch. The establishment of Nile Perch in Lake Victoria is not only on economic and ecological tragedy but a loss to the evolutionary biology (Barel *et al.*, 1985).The introduction of carnivore *Clarias gariepinus* to the Eastern Cape in South Africa has similarly devastated some native fishes and crabs.

The introduction of sea lamprey (*Petromyzon marinus*), Rainbow smelt (*Osmenus mordax*) and common carp (*Cyprinus carpio*) have affected the native species like mosquito fish (*Gambusia affinis*) and *Fundulus heterochitus* (Gracia Marian *et al.*, 1990).

Release of non-indigenous fishes in United States began as early as the 1600s with the introduction of the goldfish (*Carassius auratus*) (Crossman,1991),and by the late 1800s,non-indigenous fish introductions were also accompanied by translocation of native fishes to new locations (Carlander, 1954,Courtenay 1979, Courtenay 1990,Crossman 1991).With reference to history, Crossman (1991) stated that there were two major periods of fish introductions in North America, the late 1800s and the time following the 1950s.This resulted in 35 exotic fishes established in North America by 1980 with an additional 68 species by 1991 (Crossman 1991). Nico and Fuller (1999) and Fuller *et al.* (1999) provided the most current information on distribution of non-indigenous fishes in the United States noting that 316 native fishes have been introduced outside their normal geographic range by humans,185 species have been introduced from foreign countries, and 35 species are hybrids. The numbers of introduced species cited by Nico and Fuller (1999) and Fuller *et al.* (1999) are somewhat higher than those of past researchers because they included all fish introduced into the United States since the 1800s whether established or not, and they also count stocked fish as being introduced, even if it was moved within its normal geographic range (*i.e.*, translocated) (Neil and John, 2001).

Tilapia was first introduced in China, in 1956 from Vietnam and Africa. The tilapia industry in China is maintaining an annual growth rate of about 15 per cent. The China's ministry of Agriculture and the provincial governments are encouraging tilapia culture by providing subsidies to the farmers to the extent of US $ 300-500 per ha. The returns from tilapia culture are higher than that from other cultured fish. In view of this, many farmers, who were traditionally culturing carps, are now changing over to tilapia culture and tilapia culture is considered as a major means of raising income for rural population (Rao,2008).

Brazil has imposed strict restrictions on introducing species into its river courses. Even trans-basin movements within the country are not allowed normally. But the small reservoirs in the nine states of the northeast are the exceptions where many Chinese and African species have been exceptionally well in the small water bodies of the northeast especially in the states of Ceara and Pernambuco (Gurgel, 1984; Sugunan, 1996).

Introduction of *Oreochromis mossambicus* in Parakrama Samudra reservoir of Sri Lanka resulted into increase of fish production from 10 kg/ha/yr (in 1953) to 450-500 kg/ha/yr (in 1978).Indigenous fishes have not been negatively affected and, on the contrary, it appears that they have benefited from the presence of tilapias. It seems likely that the high densities of tilapia speed up mineralization thus enhancing plant growth. Parakrama Samudra supports a very dense fish eating bird population and also indigenous piscivorous fish species. Tilapias probably reduce predation pressure on indigenous carps. Overall, the average fish yields in Srilankan reservoir fisheries is about 300 kg/ha/yr and *Oreochromis mossambicus,* introduced in 1952,accounts forever 70 per cent of the total landings and do not compete with native species (Jerome and Lionel,2002).

Pakistan has introduced several alien exotic fish species *e.g.* grass carp (*Ctenopharyngodon idella*), bighead carp, (*Aristichthys nobilis*), silver carp, (*Hypophthalmichthys molitrix*), common carp (*Cyprinus carpio*), gold fish (*Carassius auratus*), and three species of tilapia (*Oreochromis aureus, Oreochromis mossambicus, Oreochromis niloticus*) in warm waters along with rainbow trout (*Onchorynchus mykiss*) and brown trout (*Salmo trutta fario*). The exotic species are becoming invasive in the freshwater biomes of the Punjab and other provinces of Pakistan by reason of their potent reproductive potential and feeding competitions with the native freshwater fish fauna. Resultantly the native fish species viz; *Channa marulius, Wallago attu, Rita rita, Sperata sarwari, Gibelion catla, Cirrhinus mrigala* and *Labeo rohita,* which are of economic value are under threat (Khan *et al.*2011).

In Pakistan *Oreochromis mossambicus* was introduced from Malaya in 1951 (Naik, 1973). Whereas *Oreochromis aureus,* and *O. niloticus* were introduced from Egypt during the year 1985 under 1st aquaculture development programme sponsored by Asian Development Bank. Tilapia has escaped from aquaculture and has established on its own in many wetlands in Sindh, while grass carp, introduced in the 1970s into Haleji Lake for controlling weeds, has reportedly deprived the native 74 herbivorous fish of their food. *Oreochromis aureus* is the most commonly found alien species in the Trimmu and Balloki reservoirs. The economically important fishes of Pakistan (*Labeo rohita, Cirrhinus mrigala, Gibelion catla,*) are basically vegetarian or detritivorous, where as alien fish species *Oreochromis aurius, O. mossambicus, O.niloticus* are omnivorous that feed mainly on plankton and aquatic vegetation. *O. aureus* is a generalized feeder with a preference for detritus and decanted phytoplankton. It also eats small fish and fish larvae. Due to wide ecological range alien *Oreochromis aureus* has been established in the freshwater reservoirs of Pakistan. This species is in direct competition for food to native herbivore fishes.*O.aureus* may pose serious adverse impacts in future on native fishes of Pakistan such as *Labeo rohita, Cirrhinus mrigala, Gibelion catla, Sperata sarwari, Rita rita, Channa marulius* and *Wallago attu* due to its competition for food, breeding grounds, predation and habitat alteration.

A substantial proportion of Canada's freshwater fish and molluscs are at risk for a variety of reasons. Habitat alteration, pollution and alien invasive species are the dominant threat factors influencing species at risk in Canada.

The introduction and establishment of exotic freshwater fishes in Australia has been documented (McKay,1989;Arthington and Mitchell,1986;Polland *et al.*, 1980).Brown trout feeding may have caused the decline of the Tasmania mountain shrimp, *Anaspides tasmaniae,* and has eliminated or reduced several Plecoptera and Trichoptera in Victorian streams (Fletcher,1979). *P. latipinna, X.maculatus* and *P. caudimaculatus* are much more patchily distributed in Australia, but in sub-tropical and tropical areas of Australia, these species could spread and have an adverse effect on endemic fishes (McKay,1978;Balla *et al.*1985). *Gambusia affinis,* an introduced freshwater fish in Australian waters has become a very common species which is partly responsible for the decline of several endemic fishes, including species which play an important part in keeping mosquito populations under control (Arthington *et al.*1983;Llyod 1984,1986;Lloyd *et al.*1986).Thus *G.affinis* is a pest species, or ecological 'weed' (Lloyd 1984;Lloyd *et al.*, 1986).It has had destructive effects on invertebrate populations in field situations (Stephanides 1964;Legner and Medved 1974). It attacks the eggs and fry of important sport fishes (Mayers, 1965) and also eats fish fry in aquaria (Johnson 1976;Lloyd 1987).

Since the late 19th century, a great number of exotic species of aquatic organisms have been introduced into Japan primarily with a view to enrich the fish resources of the country. While the introduction of some of these species have caused serious ecological problems in natural waters (Chiba *et al.*1989).More than 120 exotic species of aquatic animals have been brought to Japan up to 1986.Of these, only 36 species were introduced to Japan before 1945.Many of these species were introduced intentionally for stocking in the natural environment, commercial culture and/or sport fishing. Some species, such as the silver carp (*Hypophthalmichthys molitrix*) and the bitterling (*Rhodeous ocellatus*) came to Japan accidentally. These two species arrived in Japan mixed with fry of the grass carp (*Ctenopharyngodon idella*) (Nakamura, 1955).Of the 120 species brought into Japan, about 9 species accounting for 7.5 per cent of the total number have acclimatized themselves to the natural environment as self-reproducing populations; about 32 or 26.7 per cent have been successfully bred; about 26 or 24.6 per cent have disappeared; the status of about 53 or 41.2 per cent remains obscure (Chiba *et al.*, 1989). In Japan, *Rhodeus ocellatus smithi* being replaced by *R.o.ocellatus*, colonization of northeastern Japan by native born species found in southwestern Japan, and the predation of largemouth bass *Micropterus salmoides* on many native fish species in certain places (Chiba *et al.*1989).

Fish introductions in Malaysia were started in the early 19th century along with the immigration of the Southern Chinese who brought along the techniques of fish farming (Ang *et al.*1989).Shariff (1980) indicated that Lernaea was brought into the country probably through exotic ornamental fish.Intruduction of Chinese carps, Indian major carps, rainbow trout and black bass have very little visible ecological impact on the freshwater ecosystem (Ang *et al.*1989).There is no negative impact of introduced species on human life-styles, customs or economic systems. In fact, some introduced food fishes such as Chinese carps and the snakeskin gouramy have positive economic impacts as well as nutritional contribution to rural communities (Ang *et al.*1989).

The first recorded introduction of an exotic fish species to the Philippines was in 1907.The Philippine freshwater catfish, *Clarias macrocephalus* became completely dominated by an exotic species, *Clarias batrachus*, imported from Thailand during the craze for catfish farming patterned after the success in Thailand.*Clarias batrachus* now dominates natural populations in lakes and rivers and the indigenous *Clarias macrocepahuls* can hardly be found in the markets (Juliano *et al.*1989).Other introductions did not make much impact and these were not regarded by Filipinos as significant in terms of benefits and adverse effects.

The introduction of *Tilapia mossambica* was a real nightmare for brackishwater farming, competing for food in the farm with *Chanos chanos* (milkfish).*Tilapia mossambica* is now an established species in brackish water farms in the entire country (Juliano *et al.*1989).

There are 38 aquatic species known to have been introduced to Taiwan. History of introductions of exotic aquatic species into Taiwan has not been well recorded. Many of the exotic aquatic species introduced into Taiwan, such as the tilapias, have been greatly beneficial. A few, like the giant African snail, *Achatina fulica* and the apple snail, *Ampullarius insularum*, have caused serious damage to the local environment (Liao and Liu,1989).

The first record of an exotic aquatic species in Thailand can be traced back to Ayudhaya Period of about the year 1691-1692.At present there are about 15 species of fishes which were introduced for aquaculture purposes (Piyakarnchana,1989). Among them only a few are confirmed as reproducing in the natural habitats. The exotic species in the aquarium trade on the contrary were reported to increase rapidly from 69 species in 1963 to about 92 in 1988.These fish carried parasites such as anchor worm (*Lernaea* sp), cotton wool disease (*Saprolegina* sp.), and fish louse (*Argulus* sp).There are no reliable records on the degree of the destruction by these diseases to the native species. Records on the real impact to the natural ecosystem from the introduced species are scare.

Procambarus clarkia (Louisiana cryfish) has been introduced for aquaculture as well as a biological control agent for snails. It is escaped from aquaculture sites and is responsible for the disappearance of water lilies and submerged vegetation, as well as many species of snails, in the wetlands of Eastern and Southern Africa (Howard and Matindi, 2003).It threatens the existence of smaller fish, and its habit of burrowing can result in damage to dams.

Nile tilapia from Northern Africa has become an invasive alien species in Southern Africa. This species can hybridize with the Mozambique tilapia, threatening its very existence in its native habitat (Van der Waal, 2002).

Exotic and translocated fish species have become established in various parts of the inland waters of Turkey (Deniz Innal and Fu¨ sun Erk'akan). A total of 25 exotic species have been introduced. Some of these fish have been used only in closed systems while others have been released into open inland waters throughout the country. *Gambusia* sp. is possibly one of the earliest deliberate exotic introductions although there are no exact records. *Cyprinus carpio* was first cultured in Turkey at the beginning of 1960s. It successfully maintains populations from escapes and releases throughout Turkey. The mosquito fish was introduced into freshwater systems in Turkey for malaria control as an important alternative to pesticides. *Gambusia affinis*, introduced in various habitats for mosquito control, is known to be a threat to other fish. The species presence should be regarded as a serious threat to populations of *A. anatoliae* (Wildekamp and Valkenburg, 1994).

Performance of Exotic Fishes in India

The introduction of exotic fishes in Indian waters can be traced as having more than a century old history. While the country was under British rule, such fishes were possibly introduced as a means for recreational fisheries. Sir Francis Day was probably the first person who tried to introduce the Brown trout (*Salmo trutta fario*) in the Nilgiri waters in the year 1863, but this attempt was unsuccessful. In the year 1874,McIvor has introduced the trench *Tinca tinca* and the gold fish (*Carassius carassius*) into the Ooty (Ootacmund)Lake of Nilgiris. In the year 1900,the sea trout (*Salmo trutta trutta*) was brought from U.K.Later, Wilson succeeded in introducing the rainbow trout (*Oncorhynchus mykiss*) in 1906 in Jammu and Kashmir, Himachal Pradesh, Western Himalaya and Nilgiris. Further, introduction of *Salmo lavenensis* and *Onchorhynchus nerka* ended up with unsuccessful results. The giant gouramy (*Osphronemus goramy*) was first introduced in Calcutta from Java unsuccessfully during the first half of nineteenth century. In 1916, Gouramy was further introduced in Tamil Nadu from Java and Mauritius and subsequently transplanted in other states.

About 300 exotic fishes have been introduced into India for different reasons. A majority of them are ornamental fishes while some others have been introduced with the major objective of broadening the species spectrum in aquaculture, still come others have been introduced for sport and disease control.

1. Tilapia (*Tilapia mossambica*)

Tilapia is native of Africa and the middle East. They have spread mainly through introductions for fish farming and are now found in all tropical and semitropical continents. Tilapia is now being farmed in more than 90 countries. It is cultured in freshwaters of ponds and reservoirs of Java, Malaysia, Thailand, India, Pakistan, Vietnam, South Africa, Uganda, U.A.R., Zambia etc and the backwater ponds of Indonesia, Malaysia, India, Pakistan *etc.* Some tilapia species can adapt to full strength seawaters also. The nickname 'aquatic chicken' has been used on tilapia, due to its availability to grow quickly with poor quality inputs. In India, the first consignment of tilapia was brought by the Central Marine Fisheries Research Institute (CMFRI), on August of, 1952 from Bangkok and the second by the Madras Fisheries Department in the same year from Ceylon.

Figure 10.1: ***Tilapia mossambica.***

Tilapia is hardy but display some of the most undesirable characteristics. It bears at a very small size, is difficult to grow to a reasonable market size (100 gm+) with huge differences in growth between the sexes. Due to all these characteristics, tilapias are referred as 'weed fish'. The introductions of tilapia into Indian waters have not been a very happy experience. It has adversely affected the indigenous gene pool.

The presence of tilapia in carp nurseries seriously affected the survival and growth of carp fry since tilapia not only feeds extensively on carp fry but its young compete directly with carp spawn and fry for food. An experiment conducted at Cuttack during 1958-59 to determine the influence of tilapia on the survival and growth of 15 days old fry of rohu and common carp demonstrated that 1) the total fish production in ponds stocked with rohu and common carp fry was more than that of tilapia, rohu and common carp 2) The presence of tilapia population appeared to have brought in a fall in the survival of rohu fry by 17 to 22 per cent and the common carp by 16 per cent 3) In the case of rohu as well as common carp the presence of tilapia seemed to have caused very poor growth, the average individual weight attained being only about half of that in the control ponds, and, 4) In ponds where all the three species were stored together, the presence of rohu and common carp adversely affected production of tilapia to the extent at any 36 per cent as compared to controls, where the presence of tilapia adversely affected production of common carp to the extent of 61 per cent as compared to controls where the presence of tilapia adversely affected production of common carp to the extent of rohu to about 64 per cent.

In pawai lake of Mumbai the major carps have been badly hit with the accidental introduction of *O.massambicus* with the production sliding down from 33.3 kg/ha to 11.9 kg/ha (Das *et al.*, 1990).

Sreenivasan (1967) found that the growth of rates of *Catla catla, Labeo fimbratus* and *Cirrhinus mrigala* were adversely affected by tilapia in Ayyamkulam pond. He also reported that the growth of *Chanos chanos* was restricted to less than 100 gm/year, against the usual 500 gm/year in many water bodies of Tamil Nadu due to co-existence with tilapia. In Kabini reservoir tilapia has adversely affected the indigenous *Cirrhina reba. During* the period from 1980-81 to 1984-85, tilapia has caused decrease of *Cirrhina reba's* share in the catch from 70 per cent to 20 per cent (Murthy *et al.*, 1986).

The accidental introduction of tilapia in lake Jaismand (Rajasthan) has already been reported by Fishing Chimes. The catch has registered more than two fold increase in month of fishing and the whole catch was only of tilapia. The size of tilapia was found reduced compared to that of early period It is probably due to the inadequacy of food in reservoir.

Tilapias also form an important fishery of the bheries in West Bengal. Impact of tilapia in the bheries is very severe. The carp fishery that contained up to 93 per cent in 1959 was reduced to nil by 1971, associated with an overall reduction in the yield to the extent of 87.7 per cent.

According to Jhingran (1985) tilapia is unsuitable for culture along with Indian major carps because of the adverse effect it cases on the growth and production of carps and its depredations on carp fry. Although under certain condition, as under monosex culture, it can grow in to a big size, tilapia is not preferred by consumers due to its usually small size. Das *et al.* (1990) also strongly recommended banning of introduction of exotic fish species in our natural water.

Tilapia caused the deaths of more than 100 gharials in Chambal waters (Awasthi, 2008). Evidences suggest that the critically endangered reptiles could have consumed the fish abundantly found in the Yamuna which is polluted with heavy metals. The Yamuna meets the Chambal 40 km downstream of the affected area. A toxin in the fish could have entered the gharials' body and damaged their kidney, resulted in gout like symptoms and then death. Several studies have already shown that tilapia is a hardy fish and can withstand high pollution and toxins such as heavy metals tend to get accumulated in its body. The tilapia could have brought toxins from the Yamuna, where it is dominant. The tilapia link has surprised gharial experts who say gharials usually spare the invasive tilapia. They feed on local fish. But villagers in the region says illegal fishing has led to a sharp decline in local fish, forcing the gharials to target tilapia. Experts say tilapia could also harm human health and aquatic life. People in Chambal have been eating it, since few local varieties are available. Many now fear that tilapia, which is a predatory fish, may threaten fish in nearby rivers too.

When *Oreochromis mossambicus* was introduced in India in the year 1952, the country was in its early phase of inland fisheries development. The fish was thought to be an answer to our search of a species which adapted well, bred profusely, subsisted on a variety of food under diverse ecological situations, and contributed significantly to the country's fish production. Tilapia was first stocked in ponds and then in a number of reservoirs in South India. Although it was originally planned to stock them in selected reservoirs on an experimental basis.By the end of 1960s, most of the reservoirs in Tamil Nadu and those in the Palakkad and Trissur districts of Kerala were regularly stocked with tilapia with the hope of achieving quick hikes in yield rates. Somehow the fish failed to live up its promise. Its performance in ponds were discouraging from the very beginning due to early maturity, continuous breeding, overpopulation and dwarfing. In the worst cases, it was reported to mature at 6 cm length at an age of 75 days and breed at an interval of one month under tropical conditions. Performance of *O.mossmbicus* in different water bodies including the culture –based fisheries has been assessed by several authors. Menon and Chacko (1957),Menon and Krishnamoorthy (1956), and Menon *et al.*(1959) have tried to ascertain the performance of fish in respect of its feeding habits, growth and possibility of raising them as forage for the murrels under culture systems. Sreenivasan (1967) has summed up the utility of tilapia in Indian waters based on his elaborate analysis of its performance in Tamil Nadu waters. This seems to be the only serious and critical evaluation of the species in India till the 80s taking into consideration both its bio-ecology and production trends. Most other literature on the species is restricted to its yield prospects, studies on sex reversal and the physiological responses to environmental stresses. The second comprehensive review on tilapia in Indian waters was by Jhingran (1983) who analyzed in detail the pros and cons of tilapia's performance and expounding its role in the capture fishery systems of India. The committee for introduction of exotic aquatic species in India set up by the government of India has also gone into the details of the species and the possible impact of its wide –scale adoption in Indian waters for enhancing the production. Some commercial-houses have entered the field in the past by employing biotechnological approaches in tilapia farming, the most prominent among them was the venture of Vorion Chemicals, Madras which cultured the hybrid variety,the golden tilapia (Rangaswami,1988).M/S Vorion Chemicals is reported to have achieved a production rate of 65 t per hectare per year in its farm.

In Amaravathy reservoir tilapia forms about 90 per cent of the catch (Das, 2008). In Krishnagiri reservoir, tilapia production increased and tilapia contributed 68.90 per cent in the year 1989-90 with major carp fishing of 3.7 to 6.6 per cent.In Vaigai reservoir, *Oreochromis mossambicus* is the single largest component of fishery. In Pambar reservoir percentage of fish yield obtained from tilapia production was the highest. Mahanta *et al.* (2003) observed that Chullair, Meenakara, Peechi and Malampuzha reservoir in Kerala were stocked with tilapia in the early sixties and presently the fish contributes substantially in catch. In Malampuzha reservoir (2313 ha) tilapia forms 70 per cent of the catch. Tilapia also found its way into Kolleru Lake in Andhra Pradesh and Sondur reservoir of Raipur district of Chattisgarh (Kurup and Radhakrishnan,2008).Khan *et al.*(2008) noticed tilapia, *Oreochromis mossambicus* in the waters of the Chilka Lake (Orissa) where they were never introduced with a purpose. Such accidental or deliberate release of exotic fishes often takes place. In a number of Indian reservoirs, introduction of tilapia has resulted in poor growth rate or even elimination of the indigenous species and the transplanted Gangetic carps and it is certain that tilapia population is affecting the native fish fauna (Sugunan,1995).The growth rate of *Catla catla, Labeo fimbriatus, Cirrhinus mrigala* and *Chanos chanos* were adversely affected due to their co-existence with tilapia (Sreenivasan,1967).In Kabini reservoir of Karnataka, the catch of native carp species, *Cirrhinus reba* has decreased from 70 per cent of total catch in 1980-81 to mere 20 per cent in 1984-85 (Murthy *et al.*1986).

Oreochromis mossambicus is now recorded from of the polluted rivers like Ganga and Yamuna. Invasion of *O.mossambicus* in the riverine fishery has been found to aggravate the threats to the indigenous fish diversity including environmental problems.

In Wan reservoir of Maharashtra tilapia (*Oreochromis mossambicus*) affected the phenomenal growth of *Catla catla.* The growth of *Catla catla* never exceeded 850 gms after entry of tilapia. From the interview with fishermen it is learned that tilapia accidentally entered in reservoir in last five years. It may be due to other predatory fishes present in the reservoir, tilapia population could not proliferate much. Tilapia is a non-predatory fish, comes to maturity early and starts breeding, almost continuously, from the age of three months. The new recruits also multiply, compete for food and space, not only among themselves, but also with fish fauna present in the water body, which results in an over population of small sized fishes of very low/no market value. Hence, many regard tilapias as a pest and it affect adversely indigenous fish population, mainly *Catla catla.* Though this fish became a favorite of fish culturists through the world it has certain undesirable traits such as early maturity, prolific breeding and stunted growth. Being a prolific breeder, it over-populates, and very small sized tilapia are obtained due to its precocious breeding. It is therefore difficult to culture all the progeny to a marketable size and hence there is market decline, in its culture operations and is now braded as a 'trash fish'. It is not suitable for stocking with carps as it is competed with carps for space, food and oxygen besides predating on the carp fry. The accidental entry of tilapia in wan reservoir posed a great challenge for its effective control.

Invasion of *O.niloticus* has recently been reported in Yamuna river at Allahabad. The only demerit is that being a prolific breeder, it becomes a dominant species there by reducing the growth rate of the other species cultured along with it.

2. Silver Carp (*Hypophthalmichthys molitrix*)

It naturally occurs in rivers by south and Central China. It is on e of the Chinese major carp and has been introduced into Indian waters in 1959 for cultural Purpose It is a surface feeder. Its fry mainly feed on zooplankton, where as, fingerlings and adult feed primary on phytoplankton.

According to Karamchandani and Mishra (1980) silver carp and catla share a common niche and compete with each other for food in a reservoir ecosystem. Percentage composition of phytoplankton in the guts of both the fishes caught during the same time from Kulgarhi reservoir was more or less the same. Zooplankton, the favorite menu of catla, formed 21 per cent of the gut contents of silver carp. The authors concluded that silver carp hampered the growth of catla in the reservoir and advocated caution before its stocking in Indian reservoirs

In Gobindasagar reservoir, stocking of silver carp has since become irrelevant, as the fish has already carved out a nice for itself in the reservoir. In the process, the lurking fear that the exotic fish is deleterious to the population of indigenous catla has been proted beyond any doubt. Ever since silver carp gained a stronghold in the reservoir, catla which constituted an annual fishery of the magnitude of 200 to 300 t has declined considerably.

Silver carp (*Hypophthalmichthys molitrix*) is known for its ability to efficiently consume phytoplankton, which is the most dominant fraction of plankton in Rihand reservoir of Uttar Pradesh (Haque *et al.*2008).This may be just one reason of steady increase of silver carp in Rihand reservoir. But the trend may prove catastrophic for the catla fishery, which has now attained a declining status or more or less similar production over the last several years.

In Gobindsagar reservoir (Himachal Pradesh) *Catla catla* has already been replaced by the silver carp (*Hypophthalmichthys molitrix*). The main fishery of Govindsagar reservoir consists of silver carp, which fetches low price with low keeping quality and low demand.

3. Common Carp (*Cyprinus carpio*)

Its original natural distribution was probably restricted to narrow belt of Asia. It has been transplanted into many countries. In India, It was introduced in 1939. There are three varieties of *Cyprinus carpio.*

Scale carp (*Cyprinus carpio* var. *communis*): Whole body is covered with regularly arranged small scales.

Mirror carp (Cyprinus *carpio* var. *specularis*): some parts of the body are covered with irregularly arranged scales.

Leather carp (*Cyprinus carpio* var. *nudus*) Body is devoid scales.

Figure 10.2: *Cyprinus carpio.*

It is a bottom feeder, omnivorous and feeds on vegetable debris, insects, worms crustaceans and planktonic algae.

Common carp is not suitable fish for stocking in Indian reservoirs, especially the larger ones, for diverse reasons. Being a sluggish fish, its chances of survival in a predator-dominated reservoir are very poor. Due to its slow movement and bottom dwelling habit, they are not frequently caught in a passive fishing gear like gill net. It is no wonder, despite a regular stocking for 13 years, not a single common carp was ever caught from Nagargunasagar reservoirs of Andhra Pradesh. The stocked fishes failed to survive among the predators. A more important disqualification is its propensities to compete with some important indigenous carps like *Cirrhinus mrigala, C.cirrhosa* and *C. reba* with which common carp shares food niche. The presence of common carp has resulted in the decline of Cirrhinus species in Girna (Maharashtra) and krishnarajsagar reservoir (Karnataka).

Common carp is also known to compete with indigenous species such as *Cirrhinus* spp. (Dehadrai, 2008).Similarly the *Osteobrama belangiri*, the endemic fish to Loktak Lake (Manipur) is disappearing fast due to the introduction of common carp (Lakra *et al.*,2006).

Common carp caused alarming displacement and almost elimination of the prized snow trouts (*Schizothorax* spp.) from Kashmir basin by the German phenotype of common carp. The snow trout species biologically handicapped by low fecundity, and ecologically at a disadvantage by the stream breeding behavior which restricts its breeding period of snow –melt period during April-June progressively declined in population density in the face of fierce ecological composition from Common carp which established its pervasive presence all over the lake forming 75 per cent of the catch.

The chances of survival of the common carp, which is sluggish in nature, are very poor in a predator-dominated reservoir. Sugunan (1995) observed that the presence of common carp has resulted in the decline of *Cirrhinus* sp. In Girna and Krishnarajsagar the mirror carp has threatened the survival of a number of native species. In Dal Lake, the fish propagated itself profusely causing serious damage to the populations of indigenous snow trouts like *Schizothoraichthys niger*, *S.esocinus* and *S.curvifrons*.The mirror carp has caused similar damage to the snow trouts in Gobindsagar reservoir and *Osteobrama belangiri* in Loktak Lake, Manipur. Vass and Gopakumar (2002) stated that introduction of *Cyprinus carpio* is an evidence for the decline of local snow-trout fishery in some of the upland lakes in India. Recent assessment survey revealed the infestation of Yamuna river by *Cyprinus carpio* resulting in the decline of precious Indian major carp fisheries at Allahabad (Singh *et al.*, 2010). Common carp has now established itself in the Ganga river and the so colonized fishes contributed the bulk of the catch. Introduced common carp has been reported to implicate environmental changes principally eutrophication through an increase in turbidity and mobilization of nutrients to the water column from the benthos through its habit of rooting or digging in the bottom (Britton *et al.*,2007,Khanna *et al.*, 2007,Rowe 2007).The declining trend of Indian major carps in the Ganga river and increasing appearance of common carp in the fishery is a warranting situation of biological invasion threatening ecological integrity. The problem of repopulating *Cyprinus carpio* in degraded water of the river has come up to conserve rich fish genetic resources of the Ganga river before it faces a major alteration.

Feeding habit (browsing nature) of common carp could undermine reservoir banks leading to the collapse of banks and uprooting vegetation bringing changes to river courses.The foraging behaviour of common carp resulted in vegetation removal both by direct consumption and by uprooting due to its proclivity to dig through substrate in search of food.The latter activity also resulted in increased water turbidity rendering the conditions more conducive for its propagation (Lakra and Singh,2007).

The contribution of *Cyprinus carpio* in fish catch of Harni (Katgaon) reservoir (Maharashtra) was 9.33 per cent, 13.33 per cent and 13.10 per cent during first, second and third year respectively. Stocking of this species resulted in decline in catch of *Cirrhina mrigala* (Sakhare, 2015). The stocking of *Cyprinus carpio* in Gharni reservoir of Maharashtra has nearly eliminated *Cirrhinus mrigala* (Sakhare, 2015).

4. Thai Magur (*Clarias gariepinus*)

The African sharp tooth catfish, *Clarias gariepinus*, comprising 20 per cent of the total fish catch in Africa, is a remarkable and fascinating catfish species. It is extremely hardy and withstands adverse environmental conditions and habitat instability. It has been translocated to several countries from Africa and is now farmed either in its native or as a hybrid in several countries such as Republic of Korea, Peoples Republic of China, Taiwan, Philippines, Vietnam, Indonesia, Brazil, Poland, Hungary, Belgium, Bangladesh and India. It is widely known for its rapid growth and size It attains a weight of more than one kg in a year when suitable fed.

Figure 10.3: *Clarias gariepinus*.

Being highly predatory and cannibalistic in nature, the entries of the fish into Indian waters entail a serious effect on indigenous fishes. The fish has however already made an entry into Indian waters through Bangladesh and has gained wide popularity among fish farmers because of its faster growth. Unfortunately, there are highly adverse ecological implications of the proliferation of this fish and many farmers are not aware of these. The culture of the species is not therefore recommended. Studies on larval rearing of this species at Central Institute of Freshwater Aquaculture (CIFA) Bhubaneshwar indicate that even four days old spawn start chasing and injuring the others causing 2-3 per cent mortality per day as seen in the indoor rearing system.

In west Bengal fish farmers have already suffered a great loss and are ruined by the culture of this species. *C.gariepinus* is gaining popularity among fish farmers of Punjab as an economically viable alternative species to be cultured for diversification from carp culture system.

Africian catfish, *Clarias gariepinus* is native of rivers Niger and Nile and it extends to the Southern Africian rivers Limpopo, orange vaal, and Okavango and Cuene. It is now widely distributed in the world where it is native of the 40 countries while it has been introduced in more than 16 countries including India, Bangladesh and Nepal (Fish Base, 2004).

In India *Clarias gariepinus* has entered clandestinely in our culture system without any official sanction and rapidly adopted into culture practice in various parts of the country. It was first brought to West Bengal during 1993 possibly from Bangladesh from where it spread throughout the country (Thakur, 1998).In Assam it was introduced in 1994 by private fish seed traders (Baruah *et al.*, 1999).The fish has tremendous potentialities of hybridization with the local *Clarias batrachus* and *Heteropneustes fossilis*. It can also hybridize with another exotic fish *Pangasius sutchi*.The introduction of *Clarias gariepinus* has not been a success as several socio-economic aspects were neglected (Singh and Lakra, 2008).In India consumers refuse to buy *Clarias gariepinus* even at reasonable price because o f its taste, relatively large head and also because of the fact that this fish consumes slaughter house waste. The fishermen of Kolleru Lake believe that Africian catfish weighing 100 kg each are now present in the lake, and that presently the common indigenous fish species have declined/disappeared due to the invasion of *Clarias gariepinus* into the Lake (Ravindranath, 2008).It was also reported that locally available catfishes and cultured carps were badly affected due to introduction of this species in ponds and nallahs at Middle, North and South Andamans. Consequent upon this report, the Directorate of Fisheries, Andaman and Nicobar Administration took serious note of it and put complete ban over import of this live species to islands by deploying field staff at airport and seaports followed by a massive drive for eradication of the species from the culture system during 2002-2003 (Sarangi, 2008).

This species is known as a carrier of 18 species of infectious bacteria and against 3 reported in indigenous *Clarias batrachus* (Tripathi,2009).The Union Agriculture Ministry has ordered killing of these species en masse and preventing further culture of these species.

5. Bighead Carp (*Aristichthys nobilis*)

This fish has been introduced in India without official sanction. Ministry of Agriculture, Government of India, Department of Animal Husbandry and Dairying, Fisheries division, through their letter No.31016/1/96 and FY (3) dated 19th December 1997 had requested all the state governments to destroy the existing stock of magur (*Clarias gariepinus*) and big head carp (*Aristichthys nobilis*). The reason for the ban is that these may cross breed with the endemic species and cause genetic degradation of our fish fauna. National committee on exotic species had not approved the introduction of *A. nobilis* into the country. Indiscriminate stocking of big head carp in ponds had been observed to adversely affect the catla population.

Big head carp (*Aristichthys nobilis*) was introduced by CARI during 1992 in the experimental ponds. The species registered good average growth but considerable impact on catla growth was observed. The species though accepted by the consumers, the same is not recommended for the culture system to safe-guard the growth of catla (Sarangi, 2008).

6. Grass Carp (*Ctenophanyngodon idella*)

It is a natural of rivers of China. It has been introduced in India from Japan, Hong Kong in 1959 at pond culture division of Central Inland Fisheries Research Institute, Cuttack (Orissa). It has so far been rather beneficial in culture fisheries. It is believed that this trend will continue as long as the species is not able to establish itself in Indian natural waters because of easy breeding. Grass carp has very limited role-play in the natural waters. It is voracious feeder and it has proved to be very effective in the biological control of weeds like *Lemna, Hydrilla, Wolfia, Najas, Chara, Ceratophyllum, Myriophyllum etc.* By controlling weeds,it brings about circulation of the nutrients locked up in the weeds and produces valuable proteins but it impacts the survival of those fishes (murrels) and prawns that hide to escape to predators thus upsetting the ecological balance of the ponds and natural aquatic environment (Tripathi, 2009).

Figure 10.4: *Ctenopharyngodon idella.*

7. Pacu (*Piaractus brachypomus*)

It is relished as food and is also popular as aquarium fish. It is native of Amazon and Orinoco river basins of South America. It was illegally brought to India for aquarium trade and culture purpose during 2003-2004 via Bangladesh. Pacu has been reported in natural waters of Maharashtra, Andhra Pradesh, Tamil Nadu, Kerala and West Bengal. The impact of Pacu on biodiversity on long term is not properly documented. Since Pacu has a higher longevity, high fecundity and is an opportunistic omnivore, it feeds on anything, and is able to tolerate harsh environment. It can also be a competitor to the native species (Ghatge *et al.*,2014).They are known to be a traumatogenic species for humans as its strong teeth can cause serious bites (Frose and Pauly,2011).Persons engaged in capturing the fish for test sampling reported bites on their legs, thighs and on palms. Pacu has bitten off the testicles of bathing men in rivers (Smith, 2007).In many countries Pacu are legally banned for aquarium trade and aquaculture. This species is not cleared by the exotic committee in India. In spite of official ban, Pacu was brought clandestinely. Introduction of such species in freshwater would cause long term damage to our systems besides eliminating the indigenous and already existing food and ornamental fishes.

8. Gold Fish (*Carassius auratus*)

Gold fish, a native of central Europe was introduced in India as an ornamental fish. It is the most beautiful aquarium fish. More than 100 varieties of gold fish are known to exist now which show variations in body, head shapes, colour pattern and fins. Its desirable features such as relative ease in breeding, attractive appearance and immense commercial importance has made it indispensable in all aquaria.

Figure 10.5: *Carassius auratus.*

9. Guppies (*Gambusia affinis*)

It was introduced in India for eradication of mosquito. Though it is very effective as a larvicidal fish, and is destructive to native species. It has propensity to feed on fish eggs in the water bodies. It has gained importance due to its ornamental as well as public health importance.

Conclusion: Available literature on exotic fish species in Indian waters suggest that a number of exotic fish species have already entered in rivers, lakes and reservoirs, which is a matter of serious concern. There is an urgent need to educate fish farmers about the consequences of unauthorized introduction of exotic fishes and on the efforts needed for minimizing the risk factors. No exotic fish should be introduced unless it is scientifically recommended and approved. The unauthorized entry of the fish seed of undesirable species should be immediately stopped. It is duty of the extension officers/scientists to educate the farmers about the adverse consequences of culture of exotic fishes.

REFERENCES

Ambekar, E.E., Vasudevappa, Chindi and Basavaraju, Y.2008. Impact of fish introductions: A global perspective, In: Fish Introductions in India: Status, Potential and Challenges (Eds.W.S.Lakra, A.K.Singh and S. Ayyappan), pp 25-34, Narendra Publishing House, Delhi.

Ang, K.J., Gopinath, R. and Chua,T.E.1989.The status of introduced fish species in Malaysia,p71-82.In: S.S.De Silva (ed.) Exotic Aquatic Organisms in Asia. Proceedings of the Workshop on Introduction of Exotic Aquatic Organisms in Asia, Asian Fish.Soc.Spec.Publ.3,154p.Asian fisheries Society, Manila, Philippines.

Arthington, A.H.,Milton,D.A. and McKay,R.J.1983.Effects of urban development and habitat alterations, on the distribution and abundance of native and exotic freshwater fish in the Brisbabane region, Queenland. *Australian Journal of Ecology*,8: 87-101.

Awasthi, Kirtiman. 2008. Tilapia fish may have killed gharials in Chambal. *Down to Earth*, April 1-5.2008.

Balla,S.A.Liyod,L.N. and Pierce,B.E.1985.Unwanted aquarium fish,*Search*,16: 300-301.

Barel,CDN,Dorit,R.,Greenwodd,P.H.,Frayer,G.Huges, H., Jackson, PBN, Kawanabes, H., Lowe, MC Council, TH, Nogoshi, M. 1985. Destructionof fisheries in Africa's lakes,*Nature*,315: 19-20.

Baruah, U.K.; Bhagowati, A.K. and Goswami, U.C.1999.Culture of hybrid magur (*Clarias gariepinus x Clarias macrocepahlus*) in Assam, *Indian J.Fish*.46(3): 265-272.

Bijukumar,A.2000.Exotic fishes and freshwater fish diversity, *Zoos' Print Journal*, Volume XV (11): 363-367.

Britton,J.R.,Boar,R.R.,Grey,J.,Foster J.,Logonzo,J. and Harper,D.M.2007.From introduction to fishery.

Carlander, H.B.1954.A history of fish and fishing in the upper Mississipi River,Upper Mississipi River Conservation Commission,96pp.

Chiba,K.,Taki,Y.,Saki,K. and Oozeki,Y.1989.Present status of aquatic organisms introduced into Japan,p 63-70.In: S.S.De Silva (ed.) Exotic Aquatic organisms in Asia. Proceedings of the Workshop on Introduction of Exotic organisms in Asia. Asian Fisheries Society, Manila, Philippines.

Courtenay,W.R.,JR.1979.The introduction of exotic organisms, p. 237-250.In Wildlife and America. H.P. Brokaw (ed.).U.S.Fish and Wildlife Service, Forest Service and National Oceanic and Atmospheric Administration, Government Printing Office, Washington, D.C.

Courtenay,W.R.,JR.1990.Fish introduction and the enigma of introduced species. Pages 11-20.In Introduced and Traslocated fishes and their ecological effects, D.A. Pollard (ed.).Australian Bureau of National Resources. Proceeding No.8,Canberra.

Crossman,E.J.1991.Introduced freshwater fishes: A review of the North Amercian Perspective with emphasis on Canada,*Canadian Journal of Fisheries and Aquatic Sciences*, 48: 46-57.

Das, P.2008.Challenging experiences of exotic fish introductions: An overview, In: Fish Introductions in India: status,Potential and Challenges (Edited by W.S.Lakra, A.K.Singh and S.Ayyappan), Narendra Publishing House, Delhi, pp. 35-44.

Das, P., G. John and W.S. Lakra 1990. Exotic fish germplasm in Indian Reservoir P. 45 – 47. In Jhingran, Arun G. and V.K. Unnithan (eds.). Reservoir Fisheries in India. Proc of the Nat. Workshop on Reservoir fisheries, 3 - 4, January 1990 Spl. Publ. 3, Asian Fisheries society, Indian Branch, Mangalore, India.

Dehadrai,P.V.2008.Problems and prospects of exotic fishes in India, In: Fish introductions in India: Status, Potential and Challenges (Edited By W.S.Lakra, A.K.Singh and S.Ayyappan), Narendra Publishing House, pp. 17-23.

Deniz Innal and Fu¨sun Erk'akan.2006. Effects of exotic and translocated fish species in the inland waters of Turkey, *Rev Fish Biol Fisheries*.16: 39–50.

Dhawan, Asha and Kamaldeep Kaur.2001.*Clarias gariepinus* in Punjab waters. *Fishing Chimes*.21 (5): 56.

Fish Base.2004.Fish base Relational Database CD Rom.ICLARM;2005 Bloomingdak Bldg;Salceds St.legapsi village, Makati city, Metro Manila 1200,Philippines.

Fletcher, A.R.1979.Effects of *Salmo trutta* on *Galaxias olidus* and Macroinvertebrates in stream communities. M.Sc. thesis,Monash University, Victoria, Australia.

Froese, R. and Pauly, D. 2011.FishBase.World Wide Web electronic publication. www. fishbase. org,version (2/2013).

Fuller, P.L., Nico, L.G. and Williams, J.D.1999.Non indigenous fishes introduced into inland waters of the United States, American Fisheries Society, Special Publication 27, Bethesda, Mayland.

Ghatge, S.S., Belsare, S.W., Shelke, S.T. and Bopanna, B. 2014.The candidature of Pacu, an exotic fish species in unregulated aquaculture systems in the State of Maharashtra. *Fishing Chimes*. 33(10 and 11): 76-78.

Gracia-Martin,J.L.;Vila,A. ad Pla,C.1990.Genetic variation in the Iberian tooth carp,Aphanius iberus,*J.Fish.Biol*.37(Suppl-A): 233-234.

Gurgel, J.S.S.1984.Observations on the stocking of *Saratherodon niloticus* (Linne,1766) into DNOCS Public reservoirs of Northeatsern Brazil. *Bamidgeh*, 36(2): 53-57.

Howard, G.W. and Maitindi,S.W.2003.Alien invasive species in Africa's wetlands-some threats and solutions. IUCN-the world conservation Union, the Ramsar Convention on Wetlands,and the Global Invasive Species Programme. IUCN-the World Conservation Union Regional Office for Eastern Africa,Nairobi

Jerome Lazard and Lionel Dabbadie.2002.Environmental impact of introduced alien species,20.In: Encyclopedia of Life support systems (EO LSS)-Fisheries and Aquaculture: Towards Sustainable Aquatic Living resources Management, Paper available on WWW.eolss.net/sample-chapters/c10/e5-05-04-10.pdf.

Jhingran, V.G. 1985 Fish and Fisheries of India. Hindustan Publishing Corporation (India), Delhi.

Johnson,C.R.1976.Observations on growth, breeding and fry survival of *Gambusia affinis affinis* under artificial rearing conditions. Proceeding of Conference,California Mosquito Control Association.44: 48-51.

Juliano, R.O., R. Guerrero III and Ronquillo,I.1989.The introduction of exotic aquatic species in the Philippines,p.83-90.In S.S.De Silva (ed.) Exotic Aquatic Organisms in Asia. Proceedings of the Workshop on introduction of exotic aquatic organisms in Asia, Asian Fish.Soc. Spec. Publ.3,154p.Asian Fisheries Society, Manila, Philippines.

Karamchandani, S.J. and D.N. Mishra 1980. Preliminary observations on the status of silver carp in relation to catla in the culture fishery of Kulgarhi reservoir *J. Bombay Nat. Hist Soc.*, 77: 261-269.

Khanna,D.R.,Sarkar,P.,Ashutosh Gautam and R.Bhutiani.2007.Fish scales as bioindicators of water quality of river Ganga.*Environmental Monitoring and Assessment*,134(1-3): 153-160.DOI.10.1007/S1066/007-9606-5.

Lakra, W.S., Rehana Abidi, Singh, A.K., Neeraj Sood, Gaurav Rathore and T. Raja Swaminathan. 2006. Fish introductions and quarantine: Indian perspective, National Bureau of Fish Genetic Resources,Lucknow,198pp.

Lakra, W.S. and Singh, A.K. 2007. Exotic fish introduction in Indian waters, *Fishing Chimes*, 27(1): 30-34.

Legner, E.F. and Medved, R.A.1974.The native desert pupfish, *Cyprinodon maculates*, a substitute for Gambusia in mosquito control. Proceedings California Mosquito Vector Control Association,42;58-59.

Lloyd, L.N.1984.Exotic fish –useful additions or 'animal weeds'. *Fishes of Sahul*.1: 39-42.

Lloyd, L.N.1986.An alternative to insect control by mosquito fish, *Gambusia affinis*,p.156-163.In: T.D.St.George,B.H. Kay and J. Blok (eds.). Arbovirus Research in Australia, *Proceedings Fourth Symposium*, Brisbone, 1986.

Lloyd, L.N.1987. Ecology and distribution of small native fish of the lower River Murray,South Australia and their interactions with exotic mosquitofish, *Gambusia affinis holbrooki* (Girard).M.Sc. thesis, Department of Zoology, University of Adelaide.

Mack, R.N., Simberloff, D.W., Lonsdale, M., Evans, H., Clout, M., and Bazzaz,F. 2000.Biotic invasions: causes, epidemiology, global consequences and control. *Ecol.Appl*;10: 689-710.

Mahanta, P.C., R.Abidi and D.Kapoor.2003.Impact of introduction of exotic and genetically manipulated fishes in Indian reservoirs. In: Workshop on Fisheries Management in the lentic water system: stocking of reservoirs with fish seed, CIFRI Publication.

Mayers, G.S.1965.Gambusia,the fish destroyer. *Australian Zoologist*.13: 102.

McKay,R.J.1978.The exotic freshwater fishes of Queenland, Report to the Australian National Parks and wildlife Sevice,Canberra.

Menon, M.D. and Krishnamoorthy, B.K.1956.On the possible forage fish *Tilapia mossambica,* Pt.I.Its food.Madras state Fisheries Station. Reports and Year Book (1954-55).

Menon, M.D., Krishna Murthi, B.K. and Ramchandran,T.B.1959.On the possible forage fish, *Oreochromis mossambicus*, Pt.II, growth. Madras State Fisheries Station Reports and Year Book (1955-56).

Murthy, H.S., Gonda, S.S. and Ayyar, V., 1986. Fisheries of Kabini Reservoir in Karnataka, *Fishing Chimes*, 6(7): 36-38.

Nakamura, M. 1955.Kanto-heiyani shutsugenshita ishyokugo (some exotic fishes appeard at the Kanto plains).Nihon-Seiboisu-ohiri-gakkaiho 16/19: 333-337.

Neil, P. Bernstein and John R.Olso.2001.Ecological problems with Iowa's invasive and introduced fishes,*Jour.Iowa Acad.Sci*.108(4): 185-209.

Nico, L.G. and Fuller,P.L.1999.Spatial and temporal patterns of nonindigenous fish introductions in the United States, *Fisheries* 24: 16-27.

Rangaswami, G., 1988.Golden fish culture in India. NAGA, ICLARM, Manila, 11(1): 24-26.

Rao,Satyanarayana,G.P.2008.Introduction of tilapia in India: status, challenges and potential,pp.141-146, In: Fish introductions in India (status, potential and challenges) (Editors: W.S. Larka, A.K. Singh and S. Ayyapan), Narendra Publishing House, Delhi.

Ravindranath,K.2008.A note on aquaculture of exotic species in Andhra Pradesh,In: Fish Introductions in India: Status, Potential and Challenges (eds.) W.S.Lakra,A.K.Singh and S. Ayyappan), pp. 105-107,Narendra Publishing House, Delhi.

Rowe, D.K.2007.Exotic fish introductions and the decline of water clarity in small North Islands, New Zealand lakes: a multi-species problem. *Hydrobiologia*, 583(1): 345-358,DOI.10.1007/S 10750-007-646-1.

Sakhare, V.B.2015. Environmental impact of exotic fish introduction in reservoirs of Marathwada Region, Maharashtra, Final Report of Major Research Project, University Grants Commission, New Delhi.

Sarangi, N.2008. Exotic fishes and biodiversity, In: Fish Introductions in India: Status, Potential and Challenges (eds.) W.S. Lakra, A.K. Singh and S. Ayyappan), pp. 123-133, Narendra Publishing House, Delhi.

Shafland, P.L. and Lewis, W.M. 1984. Terminology associated with introduced organisms. *Fisheries*, 9: 17-18

Shariff, M.1980.Ocurence and treatment of ectoparasitic disease of aquarium fishes in Malaysia, *Malayan Veterinary Journal*,7: 48-59.

Singh, A.K. and Lakra, W.S. 2008.Africian catfish in India, In: Fish Introductions in India: status, potential and challenges (eds. W.S. Lakra, A.K. Singh and S. Ayyappan) pp 81-91, Narendra Publishing House, Delhi.

Singh, A.K., Pathak, A.K. and Lakra, W.S.2010.Invasion of an exotic fish-common carp, *Cyprinus carpio* L. (Actinopterygii: Cypriniformes: Cyprinidae) in the Ganga river, India and its impacts. *Acta Ichthyologica et Piscatoria*, 40(1): 11-19.

Smith, P.T.2007.Aquaculture in Papua New Guinea: Status of freshwater fish farming. ACIAR Monograph No.125, 124p.

Sreenivasan, A.1967.*Tilapia mossambica*: its ecology and status in Madras state, India. *Madras J.Fish.*,3: 33-39.

Stephanides, T.1964.The influence of the anti-mosquitofish, *Gambusia affinis*, on the natural fauna of Corfulakelet. Praktika Hell. Hydrobiologie Institute,9: 3-5.

Sugunan,V.V. 1995.Reservoir Fisheries of India, FAO Fish.Tech.Pap.,345: 423p

Sugunan,V.V.1996.Fisheries management of small water bodies in Africa, Asia and Latin America, FAO Fisheries Technical Paper (in press).

Thakur, N.K. 1998.A biological profile of Africian catfish, *Clarias gariepinus* and impact of its introduction in Asia. In Fish Genetics and Biodiversity Conservation (eds.) A.G.Ponniah, P. Das and S. R. Verma; Natcon Publication, Muzzaffarnagar, pp.275-292.

Tripathi, S.D.2009.Should Indian Aquaculture continue to depend on exotic species?- No. In: Aquaculture management (Eds. Goswami, U.C. and Kumar, D.), pp. 59-70, Narendra Publishing House, Delhi.

Van der Waal, B. 2002.Another fish on its way to extinction? Science in Africa.http: //www.scienceinafrica.co/za/2002/January/tilapia.htm.

Vass, K. K. and K. Gopakumar. 2002. Coldwater fisheries and research status in India; Pp. 3-29 In K. K. Vass and H. S. Raina (Ed.), Highland fisheries and aquatic resource management. India. National Research Centre on Coldwater Fisheries (ICAR), Bhimtal.

Wildekamp RH, Valkenburg K.1994.Notizen u¨ ber Zahnkarpfen–Lebensraume in Anatolien. Die Aquarien und Terrarien Zeitschrift 47(7): 447–453.

Chapter 11

Limnological Study of Jaisamand Lake (India) and its Suitability for Aquaculture and Fisheries

☆ *V.K. Balai, L.L. Sharma and N.C. Ujjania*

ABSTRACT

The aim of this study was to determine the water quality status of Lake Jaisamand and its suitability for fisheries. Water samples were collected from May 2005 to April 2006 at bimonthly interval from preselected sampling stations. The important attributes of water quality such as air temperature (31.88±1.22 °C), water temperature (24.61±0.87°C), transparency (185.56±2.38 cm), pH (8.20±0.05), electric conductivity (500.0±17.93 μScm^{-1}), dissolved oxygen (9.43±0.18 mgl^{-1}), free carbon dioxide (0.0 mgl^{-1}), carbonates (43.06±2.36 mgl^{-1}), bicarbonates (162.06±2.25 mgl^{-1}), total alkalinity (205.06±4.37 mgl^{-1}), orthophosphates (0.28±0.03 mgl^{-1}), nitrate-N (0.25±0.03mgl^{-1}) and silicate (9.96±0.31 mgl^{-1}) were observed. Water quality parameters were compared with reported optimum water quality standards prescribed for fish farming or aquaculture and found within the limit. The results revealed that aquatic environment of Jaisamand Lake is conducive for fish growth and water is not only suitable for fish farming purposes but also for irrigation and drinking purposes.

***Keywords**: Jaisamand, Fish farming, Rajasthan, Water quality.*

INTRODUCTION

Water is a precious natural resource and basic need of human thus considered as a national asset. With growing demands in various sectors it needs appropriate planning, development and management. Water quality gives a good impression of the status, productivity and sustainability of any water body for monitoring water quality which is the first step for management and conservation of aquatic ecosystems. This will also ensure conservation of its habitat by suitably maintaining the physico-chemical quality of water within the acceptable levels. The periodical changes in physico-chemical parameters like temperature, transparency, dissolved oxygen, chemical oxygen demand, nitrate, phosphate *etc.* of water may provide valuable information on its quality impacts on the productivity and biodiversity of the reservoir. Good water quality like adequate oxygen, proper temperature, transparency, limited levels of metabolites and other environmental factors are known to affect fish culture. The earlier studies on water quality of a fish farming pond in India were conducted by Sewell (1927) and after that many workers have studied the physico-chemical conditions of inland waters either in relation to fish mortality or as part of general hydrological survey (Alikunhi *et al.*, 1952 and Upadhyaya, 1964). The details of various lake ecosystems also have been studied by (Johri, 1990; Pani and Misra, 2000; Chaturbhuj *et al.*, 2004; Moundiotiya *et al.*, 2004; Sisodia and Moundiotiya, 2006; Kumar *et al.*, 2009; Mahesha and Balasubramanian, 2010 and Dubey *et al.*, 2013). In the present study, an attempt has been made to study the physico-chemical parameters of Jaisamand Lake situated in Udaipur district, Rajasthan, India to assess the various aspects of the reservoir and so as to suggest ways and means for supporting sustainable fisheries and conservation.

Material and Methods

Study Area

Jaisamand is one of the oldest manmade lake constructed during the year 1729 AD by putting an embankment across the Gomti River near the village Veerpura in Udaipur district of Rajasthan state. It covers 7,160 ha water spread area and water source for this lake are nine rivers and several seasonal cannels (Nalahas) located in the catchment area. For wide area coverage of this study three sampling stations (A, B and C) were selected (Figure 11.1).

Sampling Procedure and Laboratory Analysis

Water samples were collected in the morning hours from preselected three sampling stations during first week of each alternate month from May, 2005 to April, 2006. Water samples were collected from surface area in clean and rinsed polyethylene sampling bottles and brought to the Research Laboratory of College of Fisheries, (MPUAT) for analysis.

The important physico-chemical parameters of water including water temperature, transparency, pH, electric conductivity, dissolved oxygen, free carbon dioxide, carbonates, bicarbonates, total alkalinity, orthophosphate, Nitrate-N and silicate were analyzed following methods given by Welch (1952), Trivedi *et al.* (1987) and APHA (1989).

Results and Discussion

The range of variations and their annual mean along with standard deviation of various physico-chemical characteristics of water of Jaisamand Lake and its congeniality for fish farming are given in Table 11.1.

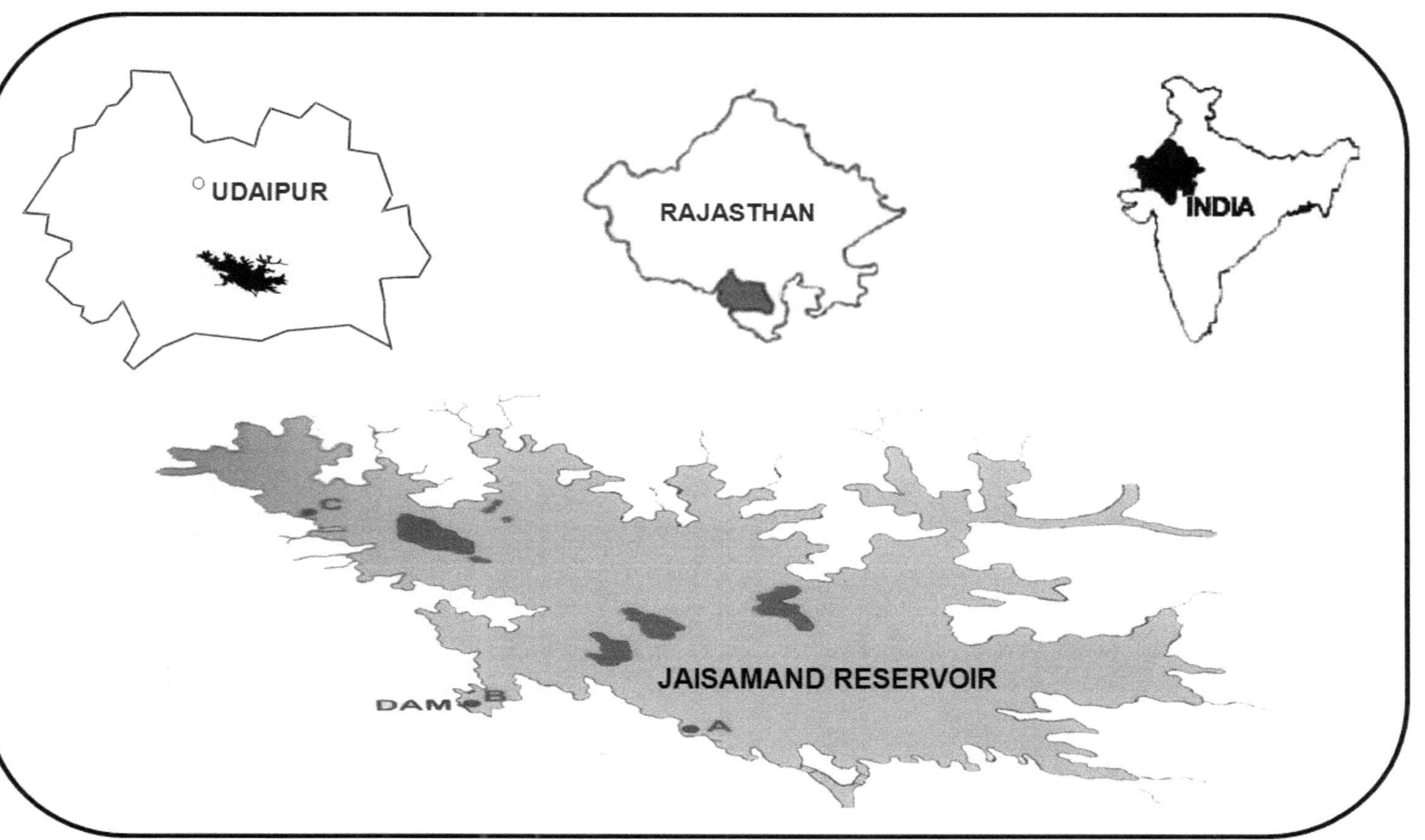

Figure 11.1:Map of Study Area in Jaisamand Lake and Sampling Stations. (A is sampling station 1, B is sampling station 2 and C is sampling)

Table 11.1: Water Quality Parameters of Jaisamand Lake

Water Quality Parameters	*Units*	*Minimum*	*Maximum*	*Mean±SE*	*Accepted Water Quality Parameters*
Air temperature	°C	24.50	41.30	31.88±1.22	–
Water temperature	°C	18.40	30.70	24.61±0.87	20-30[1]
Transparency	Cm	170.0	199.01	85.56±2.38	30-80[1]
pH		7.80	8.60	8.20±0.05	6.5-9.0[1]
Electric Conductivity	μS cm^{-1}	390.0	670.0	500.00±17.93	30-2000[2]
Dissolved oxygen	mg l^{-1}	8.50	11.00	9.43±0.18	3.0-5.0[1]
Free CO_2	mg l^{-1}	Ab	Ab	Ab	0 - 10[1]
Carbonates	mg l^{-1}	27.00	56.00	43.06±2.36	–
Bicarbonates	mg l^{-1}	147.0	179.0	162.06±2.25	–
Total alkalinity	mg l^{-1}	174.00	231.0	205.06±4.37	50-200[1]
Orthophosphates	mg l^{-1}	0.11	0.47	0.28±0.03	0.03-2.0[1]
Nitrate-nitrogen	mg l^{-1}	0.09	0.47	0.25±0.03	0-100[1]
Silicate	mg l^{-1}	8.10	11.70	9.96±0.31	2.0-20.0[3]

1: Bhatnagar and Devi (2003), 2: Stone and Thomforde (2004), 3: Marion (1998).

The temperature is an important factors which effects rate of chemical reactions of water and biological processes of aquatic organisms and thus has profound influence on the biotic communities. During the present investigation the average air temperature was 31.88±1.22 °C and water temperature was 24.61±0.87°C. Similar results were recorded by Jha and Barat (2003) in Mirik Lake, Darjeeling. Radhika *et al.* (2004) studied the abiotic parameters of a tropical freshwater lake Velayani in Thiruvananthapuram, District (Kerala) and observed the similar findings. Ujjania and Soni (2015) also reported similar findings for water temperature (25.83±0.85 °C) in Vallabhsagar reservoir (Gujarat).

Transparency is directly proportional to the amount of suspended organic and inorganic particulate matters. The other factors which affect the transparency of water body are plankton density, wind velocity, rainfall, nature of water body and prevailing weather conditions. In the present work average value of transparency was 185.56±2.38 cm. The findings of present study are in agreement with Sharma and Sarang (2004) for Jaisamand Lake (India) and Mustapha (2008) for Oyun reservoir (Nigeria), respectively.

During the present investigation average value of pH was observed 8.20±0.05. The pH of water appears to be dependent upon the relative quantities of Calcium carbonates and bicarbonates. As these being alkaline when the quantities of carbonates are high (Pearsall, 1930 and Zafar, 1996). Santhosh and Singh (2007) reported that suitable pH range is 6.7 to 9.5 for fish culture whereas above or below this level of pH is stressful to the fishes. The present value of pH is lies between the reported pH range which depicts that water condition is grossly suitable for fish and fisheries.

The electrical conductivity depicts the presence of total mineral contents in water. In the present investigation, observed average value of electric conductivity was 500.0±17.93 µScm^{-1}. Mustapha (2008) reported very low conductivity in Oyun reservoir but results of present study were found lower than the recommended values of WHO (1993) for drinking water and hence the water can be considered suitable from this point of view.

Dissolved oxygen is an important parameter which affects chemical as well as biological reactions in an ecosystem. The average dissolved oxygen during research work was 9.43±0.18 mgl^{-1}. Similar trends of DO were also observed by Yeole and Patil (2005;2007) in Yedshilake, Washim (M.S.), Ujjania (2003) in Mahi Bajaj Sagar (Rajasthan) and Ujjania and Soni (2015) in Vallabh Sagar (Gujarat).

Carbon dioxide dissolved in water is the source of carbon that can be assimilated and incorporated into the living matter of all aquatic autotrophs (Hutchinom, 1957). During the present study free carbon dioxide was totally absent. The absence of the free CO_2 may be due to its complete utilization in photosynthetic activity (Sreenivasan, 1971).The absence of CO_2 were also reported by Srivastava *etal.* (2003) in Ramgarh Lake, Jaipur, Ujjania (2003) in Mahi Bajaj Sagar reservoir, Banswara.

Carbonate and Bicarbonates were present due to absence of free CO_2 because it is converted into carbonate and bicarbonate. In the present observation mean value of carbonate and bicarbonate were 43.06±2.36 mg/l^{-1} and 162.06±2.25 mg/l^{-1} respectively. Similar trends of carbonates was reported in Chandola Lake, Karnavati (Verma *et al.*, 2012). Whereas such findings of the present study on bicarbonate is evident by Tandale and Dabhade (2014) in Lonar Crater Lake, India.

Total alkalinity is the measure of the capacity of water to neutralize a strong acid. It is generally imparted by the salts of carbonates, bicarbonates, phosphates, nitrates, borates, silicates *etc.* together with the hydroxyl ions in free-state. During the present investigation average total alkalinity was found 205.06±4.37 mg l^{-1}. Fluctuations in alkalinity might be due to alkaline particles and low production of plankton (Dash, 1993). Bhongade and Patil (2012) also reported the similar findings in Mohgavhan Lake (Maharashtra).

Average orthophosphate was 0.28±0.03 mg l^{-1} during the present investigation which is coinciding to the observations of Ahmed and Krishnamurthy (1990) in Wohar reservoir Aurangabad (Maharashtra state) and Singh and Balasingh (2011) in Kodaikanal Lake (Kerala).

In the present investigation average nitrate value was 0.25±0.03 mg l^{-1}. These values of nitrates may be due to agriculture runoff from catchment area. The finding of Ade and Vankhede (2001) in Amravati University Reservoir and Ujjania (2003) in Mahi Bajaj Sagar reservoir were similar to the present investigation.

Silicate is considered as micronutrient for the primary producers and in particular. In the present study, the mean value was 9.96±0.31 mg l^{-1} and similar findings were reported by Soni and Ujjania (2014) in Vallabhsagar reservoir (Gujarat).

Conclusion

From the present investigations, it may be inferred that physico-chemical characteristics of Lake Jaisamand water varied considerably and comparable to the other freshwater bodies. The results depict that the water of Jaisamand Lake is adequate with respect to essential nutrients necessary for primary producers which in turn is favorable and conducive for better fish growth and production. All the physico-chemical parameters in general appeared within permissible limits prescribed by different researchers. Hence, it can be inferred that the Jaisamand Lake is suitable for drinking, irrigation, pisciculture *etc.* These findings can be useful for the management and conservation of lake and its fisheries.

Acknowledgments

The Authors are thankful to the staff members of the Laboratory, Aquaculture Department, College of Fisheries, (MPUAT) Udaipur and RTADCF, Udaipur for providing the necessary facilities to complete this research task.

REFERENCES

Ade, P.P. and Wankhede, G.N. 2001. Limnological studies of Amravati University Reservoirs with reference to trophic status and conservation, Ph.D. Thesis, Amravati University, Amravati.

Ahmed, M. and Krishnamurthy, R. 1990.Hydrobiological studies of Wohar reservoir Aurangabad (Maharashtra state), India, *J. Environ. Biol.*, 11(3): 335-343.

Alikunhi, K.H., Ramachandra, V. and Chaudhuri, H. 1952. Mortality of carp fry under supersaturation of dissolved oxygen in water. *Proc. Nat. Inst. Sci. India*, 17 (4): 261-264.

APHA.1989. Standard methods for examination of water and wastewater (17th Edn.). American Public Health Association, Washington, D.C.

Bhatnagar, Anita and Devi, Pooja 2003. Water quality guidelines for the management of pond fish culture. *International Journal of Environmental Sciences*, 3(6): 1980-2009.

Bhongade, S.S. and Patil, G.P. 2012. Limnological study of Mohgavhan Lake, Karanja (Lad) District – Washim, (M.S.) India. *Vidyabharati International Interdisciplinary Research Journal*, 1(2): 27-33

Chaturbhuj, M., Sisodia, R., Kulshreshtha, M., and Bhatia, A.L. 2004. A Case study of the Jamwa ramgarrh wetland with special reference to physico-chemial properties of water and its environs, Department of Zoology, University respected y of Rajasthan, Jaipur, India Sethi Colony, Jaipur, Rajasthan, India., *Journal of Hydrobiology.*,pp 16.

Dash, B. 1993. Biological nitrogen fixation in freshwater pond ecosystem with reference to hydrobilogical conditions Ph.D. Thesis, O.U.A.T., Bhubaneshwar, p. 318.

Dubey, M., Tiwari, A.K., and Ujjania, N.C.2013. The study of physico-chemical properties of Sahapura Lake, Bhopal (India). *International Journal of Advanced Research.* 1(8): 158-164.

Hutchinson, G.E. 1957. A treatise on limnology Vol.1 Geography, physics and chemistry. New York, John Wiley and Sons, New York : 1015.

Jha, P. and Barat, S. 2003. Hydrobiological study of Mirik Lake in Darjeeling, Himalyas. *Journal Environment Biology*. 24(3): 339-344.

Johri, M. 1990. Limnological and water quality status of Bhopal Lakes with special reference to zooplankton, macrophytes and periphyton components, Ph.D. Thesis, Barkatullah University, Bhopal.

Kumar, A., Sharma, L.L. and Arey, N.C. 2009. Physico-chemical characteristics and diatoms diversity of Jawahar Sagar Lake-A wetland of Rajasthan. *Sarovar Sourabh*, 5(1): 8.

Mahesha and Balasubramanian, 2010. Analysis of water quality index (WQI) in Dalvoy Lake, Mysore city, India, Nature Environment and Pollution Technology, 9(4): 663-670.

Marion, James E. 1998. Water quality of pond, Aquaculture Research and Development Series No. 43. Alabama Agriculture Experiment Station, Aburn University, p. 5.

Mustapha, Moshood K. 2008. Assessment of the Water Quality of Oyun Reservoir, Offa, Nigeria, Using Selected Physico-chemical Parameters. *Turkish Journal of Fisheries and Aquatic Sciences*, 8: 309-319

Moundiotiya, C., R. Sisodia, M. Kulshreshtha and A. L. Bhatia, 2004. A Case Study of Jamwa Ramgarh wetland with special reference to physicochemical properties of water and its environs, *Journal of Environmental Hydrology*, 12(24): 1-7.

Pani, S. and Misra, S.M. 2000. Impact of hydraulic detention on the water quality characteristics of a tropical wetland (Lower Lake). In: *Environmental Pollution and its Management* (Ed. Pankaj Srivastava)., pp. 18-28.

Pearsall, W.H. 1930. Phytoplankton in English lakes. The proportion in the water of some dissolved substances of biological importance. *Rev. Algol.* 18: 306-320.

Radhika, C. G., Mini I. and T. Gangadevi, 2004. Studies on Abiotic parameters of a tropical freshwater lake – Vellayani Lake, Trivandrum, Kerala, *Poll. Res.* 23(1): 49-63.

Santhosh, B. and Singh, N.P. 2007. Guidelines for water quality management for fish culture in Tripura, ICAR Research Complex for NEH Region, Tripura Center, Publication no. 29.

Sewell, R.B.S. 1927. On mortality of fishes. *J. Asiat. Soc. Bengal*, 22: 177-204.

Sharma, L.L. and Sarang, N. 2004. Physico-chemical limnology and productivity of Jaisamand Lake, Udaipur (Rajasthan). *Poll. Res.* 23(1): 87-92.

Singh, R. Prathap and Balasingh, GS Regini 2011. Limnological Studies of Kodaikanal Lake (Dindugal District), in Special Reference to Phytoplankton Diversity. *Indian Journal of Fundamental and Applied Life Sciences*, 1(3): 112-118

Sisodia, R. and Moundiotiya, C. 2006. Assessment of water quality index of wetland Kalakho lake, Rajasthan, India, *Journal of Environmental Hydrology*, 14(23): 1-11.

Soni, Nandita and Ujjania, N.C. 2014. Assessment of water quality index (WQI) of Vallabhsagar reservoir, Gujarat (India). Proceeding 4th International conference on Hydrology and watershed management. Center for Water Resources, Institute of Science and Technology, JNTU, Hyderabad, 29 Oct. to 1 Nov. 2014, pp. 539-543

Sreenivasan, A. 1971. Recent trends of limnological investigations in Indian reservoirs. In workshop on all India Cocoordinated Research Project on Ecology of Freshwater Reservoirs. CIFRI, Barrackpore, India, August, 30-31.

Srivastava, N.; Agrawal, M. and Tyagi, A. 2003. Study of physico-chemical characteristics of waterbodies around Jaipur. *J. Environ. Biol.*, 24(2): 177-180.

Stone, N. M. and Thomforde H. K. 2004. Understanding Your Fish Pond Water Analysis Report. Cooperative Extension Program, University of Arkansas at Pine Bluff Aquaculture/Fisheries, USA.

Tandale, M.R. and D.S. Dabhade, D.S. 2014. The physico-chemical parameter status of Lonar craterLake, India. *Biosci. Biotech. Res. Comm.* 7(1): 50-56.

Trivedi, R.K., Goel, P.K. and Trisal, C.L. 1987. Practical methods in ecology and environmental science. Environmental Publications, Karad (India), 304 pp.

Ujjania, N.C. 2003. Comparative performance of Indian major carps (*Catla catla, Labeo rohita* and *Cirrhinus mrigala*) in southern Rajasthan. Ph.D. Thesis, Central Institute of Fisheries Education, Mumbai.

Ujjania, N.C. and Nandita Soni 2015. Assessment of water quality of Vallabhsagar reservoir (Gujarat) and its viability for fish farming. Proceedings of National Seminar on Wetlands-Present Status, Ecology and Conservation, Maharshi Dayanand College of Arts, Science and Commerce, Parel, Mumbai, August 12, pp. 246-252

Upadhyaya, M.P. 1964. Seminar on inland fisheries development in U. P. 1964; 127-135.

Verma, P., Chandawat, D., Gupta, U. and Solanki, H. (2012). Water quality analysis of an organically polluted lake by investigating different physical and chemical parameters. *International Journal of Research in Chemistry and Environment.* 2 (1): 105-111.

Welch, P.S. 1952. Limnology, McGraw Hill Book Co. New York, 538 p.

WHO 1993. Guidelines for drinking water quality. Revision of the 1984 guidelines. Final task group meeting. Geneva 21-25.

Yeole, S.M. and Patil, G.P. 2005. Physico-chemical status of Yedshi Lake in relation to water pollution. *J.Aqua. Biol.*, 20(1): 41-44.

Yeole, S.M. and Patil, G.P. 2007. Nutrient dependent hydrobiological status of Yedshi Lake, Tq. Mangrulpir, Dist. Ashim (M.S.), Ph.D.Thesis, Amravati University, Amravati.

Zafar, A.R. 1996. Liminology of Hussain Sagar Lake, Hydrabad, India.*Phykos*, 5: 115-126.

Chapter 12

Water Quality Parameters of Hiran-2 Reservoir, Gir-Somnath District, Gujarat

☆ *V.M. Chavda, V.C. Bajaniya, K.V. Tank, J.B. Solanki, P.V. Parmar and H.V. Parmar*

ABSTRACT

The present study was undertaken to assess the physico-chemical parameters of Hiran-2 reservoir of Gir-Somnath District, Gujarat. Seasonal changes in physico-chemical parameters were analyzed for a period of one year from September 2010 to August 2011.Analysis was done for water temperature, transparency, pH, conductivity, free CO_2, total alkalinity, total hardness, calcium, magnesium, chlorides, total dissolved solids, total nitrate, total phosphates, dissolved oxygen and biochemical oxygen demand so that appropriate management measures can be recommended. The reservoir found to be moderately polluted. The reservoir can be utilized for industrial, agriculture, fisheries, aquaculture and domestic use.

Keywords: *Physico-chemical parameters, Hiran-2, Reservoir, Gir-Somnath.*

INTRODUCTION

The knowledge of water quality parameters help in planning and management of water bodies for aquaculture, fisheries, agriculture and industrial uses. The healthy aquatic ecosystem is dependent on water quality parameters. Water quality monitoring is of inevitable in the conservation of water resources for fisheries, agriculture and other activities; it involves the assessment of physico-chemical parameters of water bodies. On the other hand pollutants discharged from industries, domestic waste and agriculture are responsible for deterioration of water quality parameters in the reservoirs.

The river Hiran drains with catchment area of about 518 sq. km. starting from its head waters from Gir forest and south-west direction meeting the Arabian Sea near SomnathPatan of Gir-Somnath District of Gujarat State. The catchment area of Hiran-2 project is about 349 sq. km and about 50-60 per cent area is covered by Gir forest. Hiran-2 project provide water for irrigation, municipal water supply and industries. The drainage basin Hiran river has a maximum length of 40 km with major tributaries are Saraswati river and Ambakhoi stream, and many other unknown branches make this river almost complete near Talala town.

The present study was carried out to study some of the water quality parameter of the Hiran-2 reservoir, Gir-Somnath District, Gujarat State for a period of September 2010 to August 2011. Many workers investigated the physico-chemical characteristics of reservoirs and river waters (Bade *et al.*, 2008; Garg*et al.*, 2008; Venkatesharaju*et al.* 2010; Arya *et al.*, 2011).

Material and methods

The Hiran-2 reservoir is located on 21p 2′ 33″ N latitude and 70p 28′ 43″E Longitude at village Umrethi of District Gir-Somnath district in Gujarat. Morphometric features of Hiran-2 reservoir is given in Table 12.1.

Table 12.1: Salient Features of Hiran-2 Reservoir

Location	*Village: Umrethi, Ta.: Talala, Dist: Gir-Somnath*
Lat./long.	Lat. 21p 2' 33" N and long. 70p 28' 43"E
River	Hiran
Catchment Area	349 sq. km
Mean annual rainfall	915 mm
Area at full reservoir	8 sq. km
Gross storage capacity	38.58 Mm^3
Effective storage capacity	35.0 Mm^3
Bed Rock	Basalt

Hiran-river is a major river system, flowing through western part of Gir forest supporting a variety of wildlife ecological systems and human settlements. Kamleshwar Dam, often known as Hiran-1 and Umrethi Dam known as Hiran-2 (Figures 12.1 and 12.2), are some of the major projects on the river. The water samples were collected from four selected stations – S1, S2, S3 and S4 on the Hiran-2 reservoir during September 2010 to August 2011. Water samples were collected by 2 liters blue polythene bottle, physicochemical parameters like pH, water temperature, and dissolved oxygen of the sample was determined on the spot. Electrical conductivity, total dissolved solids, biochemical oxygen demand, alkalinity, free carbon dioxide, chloride, calcium hardness, magnesium hardness, total hardness, nitrate, phosphate, sulphate are analyzed in the laboratory by following standards methods as prescribed by APHA (2005).

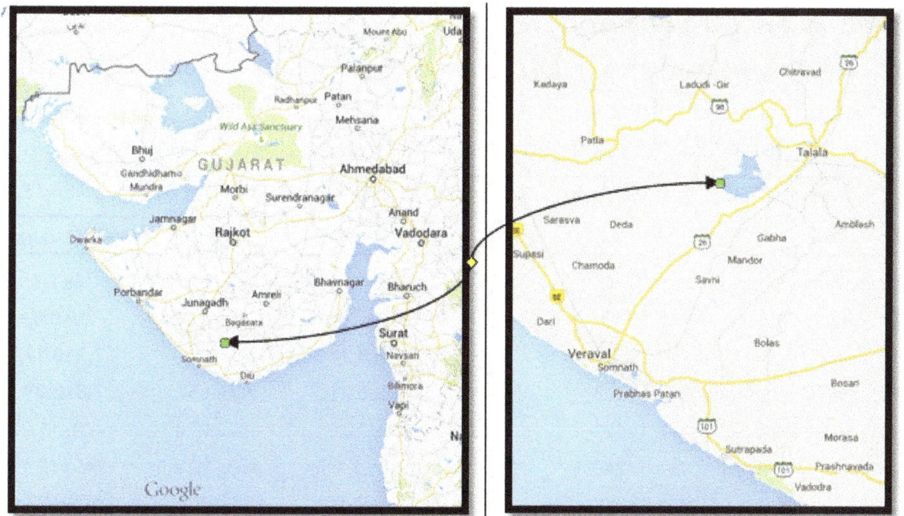

Figure 12.1: Hiran-2 Reservoir Location in Gujarat – Google Maps.

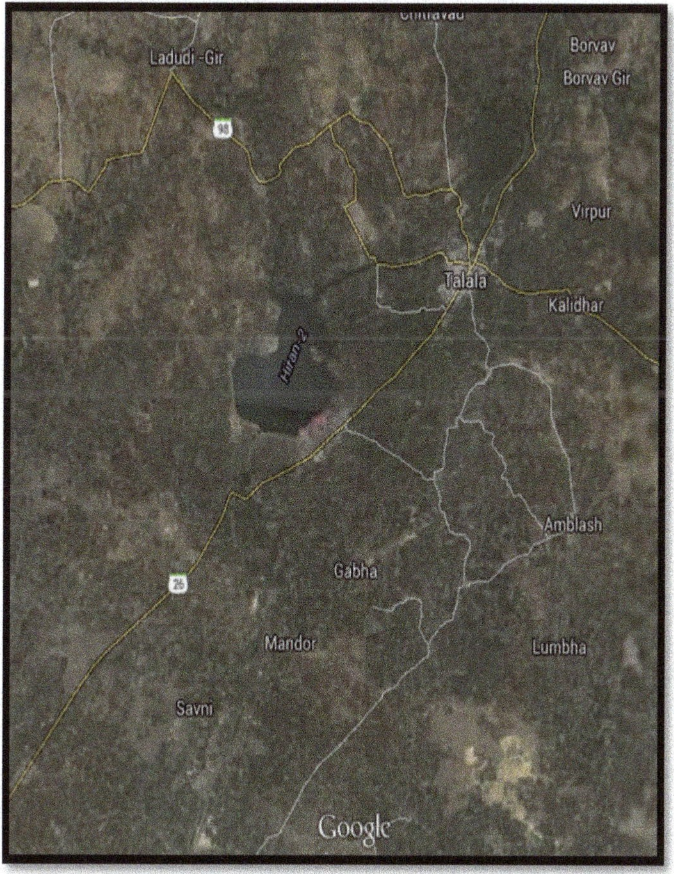

Figure 12.2: Hiran-2 Reservoir Morphology – Google Maps.

Results and Discussion

The seasonal variation of physic-chemical parameters are presented in Table 12.2.

Table 12.2: Physico-chemical Parameters of Hiran-2 Reservoir

Parameter	*Minima*	*Maxima*	*Mean Values of Stations S1, S2, S3 and S4*		
			Winter	*Summer*	*Monsoon*
Transparency (cm)	32	69	54.0±1.5	46±1.2	36±1.7
Temperature (°c)	21	28	22±0.5	27±0.4	23±0.6
pH	6.9	8.9	8.4±0.21	8.7±0.11	7.9±0.17
Dissolved Oxygen (mg/l)	6.2	8.9	8.2±0.21	7.2±0.57	7.8±0.87
Total Dissolved Solid (mg/l)	421	552	423±10.7	534±12.4	470±11.2
Conductivity (μs/cm)	249	523	297±7.2	453±12.3	240±10.4
Alkalinity (mg/l)	80	150	92±4.1	139±5.2	109±7.7
Total Hardness (mg/l)	70	140	80 ±6.5	110±11.8	93±7.4
Calcium (mg/l)	26.3	38.7	28.2±3.4	32.5±4.2	36.7±5.3
Magnesium (mg/l)	6.9	9.2	7.2±0.2	7.4±0.3	8.3±0.5
Nitrate (mg/l)	0.012	0.43	0.019±0.0012	0.29±0.031	0.3±0.027
Phosphate	0.46	0.54	0.47±0.01	0.49±0.03	0.52±0.04
Sulphates (mg/l)	8.6	18.3	9.3±0.45	13.2±1.13	11.7±1.34
Chloride (mg/l)	11.6	26.2	14.4±2.11	21.3±2.32	19.9±2.71
BOD (mg/l)	3.2	4.1	3.62±0.18	3.92±0.11	3.52±0.17
Free CO_2 (mg/l)	1.1	4.3	1.2±0.67	3.7±0.87	2.9±0.94

Transparency

Various factors like suspended solid particles, type of soil of catchment area, plankton biomass, intensity of sunlight, *etc.* determine the transparency of water body. During the present study transparency value ranged from 32 to 69 cm with sacchi disk. Highest mean value of transparency was 54 cm in the winter and lowest mean value of 36 cm was recorded during monsoon. Similar results were reported by Kadam *et al.* (2007) and Garg *et al.*(2008).

Temperature

The temperature is an important physico-chemical parameter as it regulates the life in water body (Welch, 1952) as well as it affects a number of water quality parameters. The temperature of Hiran-2 reservoir varied between 21 to 28°C in the present study. Higher average temperature values of 27°C were observed during the summer and lower average values of 22°C during winter were recorded. Sharma *et al.* (2000), Dwivedi and Pandey (2002), Singh and Mathur (2005), Salve and Hiware (2008), Garg *et. al.* (2009), Meenashi Saxena (2012), Bhadja and Vaghela (2013) and Patil (2014), found similar results.

pH

The pH plays a significant role in productivity of the water body. Hiran-2 reservoir was found to be alkaline in nature and a wide range of pH observed during the study period with values ranging from 6.9 to 8.9. Higher average values of pH was found during summer season, may be due to increased photosynthetic activity and bacterial decomposition of organic matter. Kanan and Job (1979), Jadhav and Deshmukh (2006), Jindal and Gusain (2007), Mishra *et al.* (2008), Sinha and Biswas (2011) and Saxena and Saxena(2012) have found similar results in different water bodies.

Dissolved Oxygen (DO)

DO observed at Hiran-2 reservoir was in the range 6.2 to 8.9 during the study period. Dissolved oxygen in water body is result of wind action, vertical circulation of water body as well as photosynthesis of plant matter. Minimum mean value of 7.2 was reported for DO from Hiran-2 during summer might be due to increased rate of decomposition of organic matter, and rise in temperature while maximum mean values of 8.2 was observed during winter followed by monsoon due to low temperature of winter and vertical circulation brought about by wind action and fluviatile condition of reservoir during monsoon. Similar results were obtained by Sahu*et al.* (2000), Naz and Turkmen (2005) and Singh and Singh (2011).

Total Dissolved Solids

The desirable limit of Total dissolved solids prescribed by ISI is 500 mg/L in drinking water. The concentration of TDS from natural water sources have been found to vary from less than 30 to 6000 mg/l depending on the solubility of minerals in the water bodies (WHO). The values observed during the present study period was 421 to 552 mg/l. Maximum mean values were observed during the summer followed by monsoon season. Higher values of summer might be due to higher rate of evaporation and lower water level of the reservoir. Similar results were observed by Rao *et al.* (2003), Kirubavathi *et al.* (2005), Garg*et al.* (2006), Mohammad*et al.* (2015).

Conductivity

The conductivity of water depends upon ions present in water. The electrical conductivity of Hiran-2 reservoir ranged between 249 to 523µs/cm. Higher value of conductivity can be harmful to soil structure and can cause salinity hazard on the water body. Low values of 240 µs/cm electrical conductivity of monsoon in the present investigation corresponds dilution of water during rainy season which causes decrease in electrical conductance due to addition of rain water in the reservoir whereas higher mean value of 453 µs/cm of electrical conductivity in summer may be attributed to decrease in the water level, higher temperature and high rate of evaporation and resultant increase in decomposition rate of organic matter in the reservoir. Natural waters usually have EC values of 20 to 1500 *µmhos* cm^{-1} (Boyd, 1978). However, the standard limit set by WHO is 250 *µmhos*/cm. Similar observation was made by Das (2000), Sulabha and Prakasam (2006). Krishnamurthy *et al.* (2010), Sharma *et al.* (2011).

Total Alkalinity

It is the sum total of carbonates and bicarbonates, a measure of capacity of water to neutralize strong acid. Total alkalinity is important parameter determining the productivity of the water body. Water with alkalinities of more than 50 ppm are most productive of fish, whereas waters with alkalinity of less than 10 ppm rarely produce large crops, The present investigation revealed the wide range of values from 80 to 150 mg/l. The maximum permissible limit of alkalinity is 200 mg/l for drinking water (WHO). Higher mean values of alkalinities 139 mg/l were observed during summer in the present study. Similar observations were made by Nair and Rajendra (2000), Bade *et al.* (2009), Verma*et al.* (2012) and Saxena (2012).

Total Hardness

Total hardness of water body is mainly governed by the content of calcium and magnesium salts. The observed total hardness values ranged between 70 to 140 mg/l indicates the Hiran-2 reservoir is moderately hard reservoir. Higher mean values of hardness was observed during summer followed by monsoon season from Hiran-2 reservoir. Similar behavior of total hardness was recorded by Khan *et al.* (1986), Khadade and Mule (2003), Shyam Sunder and Khatri (2015) The calcium and magnesium concentration in the present study observed from 26.3 to 38.7 mg/l and 6.9 to 9.2 mg/l respectively.

Nitrate

Organic pollution in the water body can be judged by Nitrate-Nitrogen. The nitrates are an important source of nitrogen for phytoplankton. Nitrate represents the highest oxidized form of nitrogen. The values of nitrate ranged between 0.012 to 0.43 mg/l in the current study.

Phosphates

Phosphates are essential nutrients for plant growth in water body (Jhingran, 1982) but in higher concentration causes eutrophication. Through the natural process of weathering, the rocks gradually release the phosphorus as phosphate ions which are soluble in water. In the present investigation the maximum values of 0.52 mg/l of phosphates were found during monsoon. This is in agreement with the Khan and Parveen (2012)

Sulfates

Sewage, industrial discharge and agriculture run-off contributes sulfates to water bodies. Sulfur is an essential plant nutrient as sulfates less than 0.5 mg/l algal growth will not occur. On the other hand sulfate salts can be major contaminants in natural waters. In the present study the sulfate concentration was found 8.6 to 18.3 mg/l. Sulfates in water to be used for certain industrial processes such as sugar production and concrete manufacturing must be reduced below 20 mg/l. Recommended limits for water used as a domestic water supply are below 250 mg/l. In natural water, sulphate is the second most common anion, being from most sedimentary rocks. Sulphate content was found to be more during summer

13.2 mg/l and less during winter 9.3 mg/l in the present study. Similar results were obtained by Shyam Sundar and Khatri (2015).

Chloride

The status of chloride in the water body is indicative of its degree of pollution especially due to organic waste of animal origin. The Chloride values of Hiran-2 reservoir remain between 11.6 to 26.2 mg/l, which shows moderate level of eutrophy and pollution. Higher mean values of chloride was observed during the summer. Similar results were reported by Sinha and Biswas (2011), Verma *et al.* (2012), and Lianthuamluaia *et al.* (2013) in different water bodies.

Biochemical Oxygen Demand

BOD is important parameter in estimating the pollution status of sewage and domestic waste. It indicates the presence of biodegradable organic matter quantitatively, which consumes dissolved oxygen from the water. The higher values of BOD produce obnoxious smell and unhealthy environment. In the present study the BOD values ranged from 3.2 to 4.1 mg/l. Desirable limit for BOD is 4.0 mg/l and permissible limit is 6.0 mg/l according to Indian standards. Present study reveals the Hiran-2 reservoir with lightly polluted water. Higher BOD values were observed during the summer months due to elevated microbial activity at higher temperature. Similar results were observed by Sachidanandamurthy and Yajurvedi (2004) Devaraju *et al.* (2005), Garg *et al.* (2009).

Free Carbon Dioxide

Free carbon dioxide values ranged from 1.1 to 4.3 mg/l. Present study revealed that summer season showed moderately high free carbon dioxide value. Higher values found during summer months may be due to high rate bacterial decomposition of organic matter. The results are conforming to Negi *et al.* (2006) and Bharamal and Korgaonkar (2014). As per Ellis (1937), values of free carbon dioxide values less than 5 mg/l suitable for flourishing fish populations in a water body. Present investigation suggests that the Hiran-2 reservoir is suitable for fisheries and cage aquaculture practice.

Conclusion

This investigation reveals that water quality of Hiran-2 reservoir experiences seasonal fluctuations of physico-chemical parameters. The data of present study clearly reflect that the water of Hiran-2 reservoir is moderately polluted. It is recommended that the water of Hiran-2 reservoir should be treated before its use as potable water. The water of Hiran-2 reservoir is suitable for fisheries, agriculture, Industrial as well as domestic uses.

REFERENCES

APHA, AWWA, WEF.2005. Standard methods for the examination of water and wastewater. 21st edition, Washington DC, USA.

Arya, S., Kumar, V., Sonkar, P., Minaskshi, Dhaka, A., and Chanchal, 2011. Water quality status of Historical Antiya Tal at Jhansi city as a primary data for sustainable approach. *RRST* 3(8): 52-55.

Boyd, C.E.1978. Water Quality in Warm Water Fish Ponds. Agricultural Experiment station, Auburn University, pp. 359.

Bade, B. B., Kulkarni, D. A. and Kumbhar, A. C. 2009. Studies on physico-chemical parameters in Sai Reservoir, LaturDist, Maharashtra *SSM*.Vol.II(7): 31-34.

BIS 2010. Specification for drinking water. Indian standard institution (Indian Bureau of Standard), New Delhi.

Bhadja P and VaghelaAshokkumar 2013. Hydro biological studies on freshwater reservoir of Saurashtra, Gujarat, *India. J. Biol. Earth Sci.*, 3(2): E12-E17.

Bharamal, D. L. and Korgaonkar,D.S. 2014. A preliminary investigation on water quality of tillari dam, Dodamarg, Sindhudurg, Maharashtra, *India. Int. J. Curr. Microbiol. App. Sci.* 3(7): 369-377.

Das, A. K. 2000. Limnochemistry of some Andhra Pradesh Reservoirs, *J. Inland Fish. Soc.Ind.*, 32 (2), 37- 44.

Dwivedi, B. K. and Pandey, G.C. 2002. Physico-chemical factors and algal diversity of two ponds, (Girija Kund and Maqubara pond), Faizabad. *Poll.Res.* 21, 361-370.

Devaraju, T.M., M.G. Venkatesha and S. Singh 2005.Studies on physico-chemical parameters of Muddurlake with reference to suitability for aquaculture. *Nat. Environ. Pollut. Tech.*, 4, 287-290.

Ellis, M.M.1937. Detectionand measurement of Stream Pollution. Bull. 22, US Bureau of Fisheries.

Garg, R. K., Rao R. J. and Saksena D. N., 2006. Studies on nutrients and trophic status of Ramsagar reservoir, Datia, Madhya Pradesh. *Nature Environment and Pollution Technology* 5(4): 545-551.

Garg, R. K., Rao, R. J., and Saksena, D.N. 2008. Water quality and conservation management of Ramasagar reservoir, Datia, Madhya Pradesh. *J. Environ. Biol.*, 30 (5): 909- 916.

Garg, R. K., R. J. Rao and D. N. Saksena 2009. Water quality and conservation management of Ramsagar reservoir, Datia, Madhya Pradesh. *J. of Environmental Biology*. 30(5): 909-916.

Jhingran, V. G., 1982. Fish and Fisheries of India. Hindustan Publishing Corp, (India) Delhi.

Jadhav, A.R. and A.M. Deshmukh, 2006. Physico-chemical and microbial characteristics of Rankala and Aalamba of Kolhapur district, Maharashtra, India. *Environment and Ecology*, 24(1): 21- 27.

Jindal, S. and D. Gusain, 2007. Correlation between water quality parameters and phytoplankton of Bicherli pond, Beawar, Rajasthan. *J. Aqua. Biol.*, 22(2): 13-20.

Kannan, V. and Job, S.V., 1979. Diurnal depth-wise and seasonal changes of physico-chemical factors in Sathiarreservoir,*Hydrobiologia*, 70(1-2), 103-117.

Khan, M., Raza S.A., Iqbal S.A., Shastri T. and Hussain, 1986.Limnochemistry and water quality aspects of Upper lake Bhopal during winter season, *Indian Journal of Applied and Pure Biology*, 47-50.

Khabade, S.A. and M.B. Mule. 2003. Studies on physico-chemical parameters of Pundi water reservoir from TasgaonTahsil. *I. J. Environ Prot.*, 23(9): 1003-1007.

Kirubavathy AK, Binukumari S, Mariamma N, Rajammal T., 2005. Assessement of water quality of Orthupalayam reservoir, Erode District, Tamil Nadu, *Journal of Ecophysiology and Occupational Health*, 5: 53-54.

Kadam, M. S., Pampatwar, D. V. and Mali, R. P. 2007. Seasonal variations in different physico-chemical characteristics in Mosoli reservoir of Parbhani district, Maharashtra, *Journal of Aquatic Biology*, 22(1): 110-112.

Krishnamurthy, A. and Selvaumar,S.2010. Seasonal variations in physico-chemical characteristics of water bodies in and around Cuddalore District, Tamil Nadu. *Nat. Envi. and Poll. Tech.* 9 (1): 89-92.

Khan, M. A. and Parveen, M.2012. Seasonal variations in physic-chemical characteristics of Pahuj reservoir, District Jhansi, Bundelkhand region, Central India. *International Journal of Curent Research* 4(12): 115 -118.

Lianthuamluaia, Asha T. Landge, C. S. Purushothaman, GeetanjaliDeshmikhe and Karankumar K. Ramteke, 2013. Assessment of seasonal variations of water quality parameters of savitri reservoir, Poladpur, Raigad District, Maharashtra. *The Bioscan:* 1337-1342.

Mishra, R.R., Rath B. and Thatoi H., 2008. Water Quality Assessment of Aquaculture Ponds Located in Bhitarkanika Mangrove Ecosystem, Orissa, India. *Turkish Journal of Fisheries and Aquatic Sciences,* 8: 71-77.

Mohammad, M.J., P. V. Krishna, O.A.Lamma and Shabbar Khan 2015. Analysis of water quality using limnological studies of Wyra reservoir, Khammam district, Telangana.*India. Int. J. Curr. Microbiol. App. Sci.* 4(2): 880-895.

MeenakshiSaxena, Saksena D.N., 2012. Water quality and tropic status of Raipur reservoir in Gwalior, Madhya Pradesh. *Journal of Natural Sciences Research.* 2(8): 82-96.

Nair, M. S. and Rajendran, 2000. Seasonal variations of physico-chemical factors and its impact on the ecology of a village pond at Imala (Vidisha). *J. Ecobiol.* 12(1): 21-27.

Naz, M. and M. Turkmen, 2005. Phytoplankton Biomass and Species Composition of Lake Golbasi (Hatay-Turkey). *Turk. J. Biol.*, 29: 49 56.

Negi, R. K., Johal, M.S. and Negi, T., 2006. Study of the physic-chemical parameters of water of Pongdam Reservoir, Himachal Pradesh: A Ramsar Site, India.*Him.J. Env.Zool.*20(2): 247-251.

Patil, Alaka A.,2014.Limnological and correlation studies of Birnal water body of Sangli, Maharashtra. *International Research Journal of Environment Sciences.*, 3(9): 43-49.

Rao, K.D.S, Ramakrishniah M., Karthikeyan M., and Sukumaran P.K.2003. Limnology and fish yield enhancement Reservoir (Cauvery River System), *Journal of Inland Fisheries Society of India*, 35: 20-27.

Sahu, B.K., R.J. Rao, S.K. Behara and R.K. Pandit, 2000. Effect of pollutions on the dissolved oxygen concentration pf river Ganga at Kanpur. In: Pollution and Biomonitoring of Indian Rivers (Ed: R.K. Trivedy) ABD Publication, Jaipur, India, 168-170.

Sharma, M. S. Liyaquat, F., Barbar, D. and Chisty, N., 2000. Biodiversity of Freshwater zooplankton in relation to heavy metal pollution. *Poll. Res*, 19(1): 147-157.

Sachidanandamurthy, K.L. and H.N. Yajurvedi, 2004. Monthly variations in water quality parameters (physico-chemical) of a perennial lake in Maysore city. *Indian Hydrobiol.*, 7: 217-228.

Singh, R. P. and Mathur, P., 2005.Investigation of variations in physico-chemical characteristics of a freshwater reservoir of Ajmer city, Rajasthan. *Ind. J.Env. Sci.*, 9: 57-61.

Sulabha, V. and V.R. Prakasam, 2006. Limnological features of Thirumullavaram temple pond of Kollam municipality, Kerala. *Journal of Environmental Biology*, 27(2): 449-451.

Savle, V. B.and Hiware, C.J., 2008. Study on water quality of Wanparakalpa Reservoir Nagpur, Near ParliVaijnath, District Beed. Marathwada region, *J. Aqua. Biol.*, 21(2): 113-117.

Sharma Riddhi, Sharma Vipul, ShamaMadhuSudhan, VermaBhoopendra Kumar, ModiRachana and Gaur Kuldeep Singh, 2011. Studies on limnologicalcharecaristics, planktonic diversity and Fishes (species) in Pichhola Lake, Udaipur, Rajasthan, India. *Universal J. of Environmental Research and Technology*, 1 (3): 274-285.

Singh,J. and Singh,K., 2011.Physico-chemical characteristics of Khoh River in Pauri (Uttarkand) and Bijnor (U.P.) Districts, *India. J. Env. Bio-Sci.* 25(1): 123-126.

Sinha, S. N. and Biswas, M., 2011. Analysis of physico-chemical characteristics to study the water quality of a lake in Kalyani, West Bengal. *Asian J. Exp. Biol. Sci.* 2: 18-22.

Saxena, M. and Saksena, D. N., 2012. Water quality and trophic status of Raipur reservoir in Gwalior, Madhya Pradesh. *J. Nat. Sci. Res.* 2: 82-96.

Shyam, Sunder and Khatri, A. K., 2015. Physico-chemical properties of water of Ottu Reservoir in District Sirsa, Haryana, *India. Res. J. Recent. Sci.*, 4: 190-196.

Umavathi, S., Longakumar, K and Subhashini, 2007. Studies on the nutrient content of Sulur pond in Coimbatore, Tamil Nadu. *Ecology and Environmental Conservation*, 13(5): 501-504.

Venkatesharaju, K., Ravikumar, P., Somashekar, R.K. and Prakash, K.L.,2010. Physico-chemical and Bacteriological investigation on the river Cauvery of Kollegal Stretch in Karnataka, *Journal of Science Engineering and* Technology, 6 (1): 50-59.

Verma, P., Chandawat D., Gupta U. and Solanki H., 2012. Water quality analysis of an organically polluted lake by investigating different physical and chemical parameters. *International Journal of Research in Chemistry and Environment* 2(1): 105-111.

Welch, P.S., 1952. Limnology, 2nd edition, McGraw Hill Book Co. New York.

WHO/UNEP, GEMS, 1989. Global freshwater quality. Oxford, Alden Press.

WHO, 2011. Guideline for drinking-water quality. 4th edn. World Health Organization, Geneva.

Chapter 13

A Study on Limnological Characteristics of Ponds for Composite Fish Culture for Improvement of Livelihood of Fishermen in Khurda District (Odisha)

☆ *Sasmita Panda, G.K. Panigrahi and S.N. Padhi*

ABSTRACT

Physical, chemical and biological parameters like temperature, total alkalinity, pH, dissolved oxygen, nitrate nitrogen and planktons were studied in three ponds (1,2 and 3) in Khurda district of Odisha. The parameters studied in all the ponds varied but statistically insignificant. The ponds were stabilized empowering the people with the techniques for composite fish farming using Catla, Rohu and Mrigal as candidates for harvesting.

Keywords: *Limnology, Fish culture, Khurda district, Odisha.*

INTRODUCTION

Limnology is defined as the study of freshwater and their inhabitants. The growth and survival of aquatic inhabitants depend on the quality of water (Boyd 1989,1990, Philips 1991, Jhingran 1985). The quality of water depends on the physical, chemical and biological characteristics of water (Zweig *et al.*, 1999, Adeniji and Ovie 1982; Das and Padhi 2014, Padhi *et al.*, 2015). Hence the present study is focused on the determination of quality of water in order to utilize the ponds for aquaculture. Fish plays an important role in agriculture sector of India. It provides livelihood to more than 60 million people and earns more than 6800 crore rupees through export. Limnological studies have been carried out by Olopade (2013) and Nikolosky (1963).

The main objectives of the study was to study physical, chemical and biological characteristics of ponds in order to utilize them for fish culture and thus generate employment opportunity for gainful earning among rural people by creating awareness among them through training on aquaculture practices. Composite fish farming is the technique to culture different types of compatible and non competitive fishes in the same ecosystem so as to allow them to grow by feeding by making optimum use of different zones (surface, bottom and column) of the ponds without impeding the growth, development and maturity of one another. This is a very profitable method of aquaculture and hence importance has been laid to train the populace for gainful employment.

Materials and Methods

For the purpose of study, three ponds in three villages (Khudpur, Podopoda, and Barapada) in Khurda district, Odisha, were chosen for investigation, and such ponds were not utilized for fish cultivation earlier. The parameters chosen were water temperature, pH, dissolved oxygen, total alkalinity, nitrate nitrogen and plankton biomass of water.Temperature was recorded using ordinary thermometer (accurate up to 0.01 degree C),pH by pH meter, alkalinity by using phenolphthalein and methyl orange indicators. Dissolved oxygen was measured by Winkler's method and nitrate nitrogen and plankton biomass were measured by following standard procedures (APHA-1989) using water testing kits (NICE). The study was conducted for the period of one year *i.e.* from November 2011 to October 2012.

Preparation of Ponds for Composite Fish Culture

1. Cleaned the ponds by making free from undesirable plants and weeds manually.
2. Liming the ponds was done in order to correct the acidity of soil and water to speed up the decomposition of organic matter; which acts as disinfectant and as an essential nutrient (@ 200 kg/ha).
3. Fertilizing the ponds was done after 3 days of liming for bloom of phytoplankton and growth of zoo plankton by manuring with organic manures (cowdung and oil cake @ 500 kg/ha) which carry almost all nutrients required by fish. The inorganic manures like urea @ 60 kg/ha/month and single super phosphate @ 70 kg/ha/month were used depending on the soil and water condition of the different ponds as they provide the nutrients, vitamins and minerals to the fish thus increasing natural productivity of the ponds.
4. Artificial feeding was done by providing rice bran, oil cake and kitchen waste as these are cheaply available.

5. After cleaning, liming and fertilizing the ponds during March-April 2013, the fingerlings of 50-100 gm size (approx) purchased from govt. hatcheries were stocked in the ponds 15 days after fertilization of ponds. Fingerlings of *Catla, Rohu* amd *Mrigal* in the ratio of 4:3:3 were selected to get good yield in mixed farming during July 2013.

Results and Discussion

The maintenance of good water quality is essential for both survival and optimum growth (Gupta and Gupta 2006).The water quality standards vary significantly due to different environmental conditions, ecosystem and intended human users EPA 2006. The quality of aquaculture products and their suitability for human consumption may also be affected by water quality Zweig *et al.*, 1999. Keeping these factors in view, the ponds under study were maintained for aquaculture imparting training to local people also in order to empower them for gainful employment.

The results (Table 13.1 and Figures 13.1–13.5) indicate that the water temperature in pond 1 varied from 19 to 32, in pond 2 from 19.3 to 33.5 and in pond 3 from 21 to 33.1. pH in pond 1 varied from 7.4 to 8.1, in pond 2 from 7.5 to 8.2 and in pond 3 from 7.4 to 8.1. Total alkalinity showed 81.3 to 184.86 in pond 1, from 86.20 to 196.15 in pond 3 and varied from 84.18 to 196.24. Dissolved oxygen values varied from 5.4 to 8.5 in pond 1, from 5.5 to 8.3 in pond 2 and from 5.6 to 8.1 in pond 3.Nitrate nitrogen varied from 0.98 to 5.28 in pond 1, from 0.96 to 8.21 in pond 2 and from 2.38 to 15.35 in pond 3.Phytoplankton analysis revealed variation from 212 to 218 in pond 1, 224 to 231 in pond 2 and 225 to 238 in pond 3.Zooplankton varied from 20 to 38 in pond 1, 27 to 35 in pond 2 and from 26 to 31 in pond 3 during different months of the year (2011-2012) under study. Taking all these factors into consideration, the three ponds under study were prepared for fish culture by cleaning, liming, fertilizing, stocking and artificial feeding for harvesting after one year.

Table 13.1: Range of Variation of Parameters (November 2011 to October 2012)

Parameters	*P1*	*P2*	*P3*
Temperature	19-32	19.3-33.5	21-33.1
Total alkalinity	81.36-184.86	86.20-196.15	84.18-196.24
pH	7.4-8.1	7.5-8.2	7.4-8.1
Dissolved Oxygen	5.4-8.5	5.5-8.3	5.6-8.1
Phytoplankton (no./lit)	212-218	224-231	225-238
Zooplankton (no./lit)	20-38	27-35	26-31
Nitrate nitrogen	0.98-5.28	0.96-8.21	2.38-15.35

The result (range of variation of different parameters) obtained on three ponds (P1, P2, P3, Average Depth of 3.5-5′) during the period of study(Nov 2011 to Oct 2012), were recorded (Table 13.1). After having studied, the pond health, treatment was done preparing the ponds for aquaculture following standard prescribed guidelines for pre stocking, stocking and harvesting.

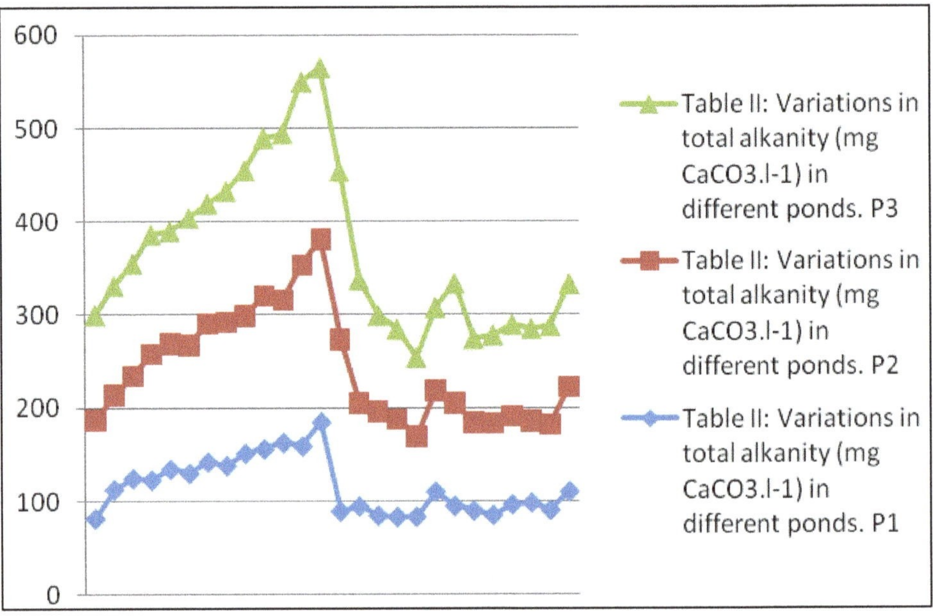

Figure 13.1: Variations in Total Alkalinity (mg CaCO3 l^{-1}) in different Ponds.

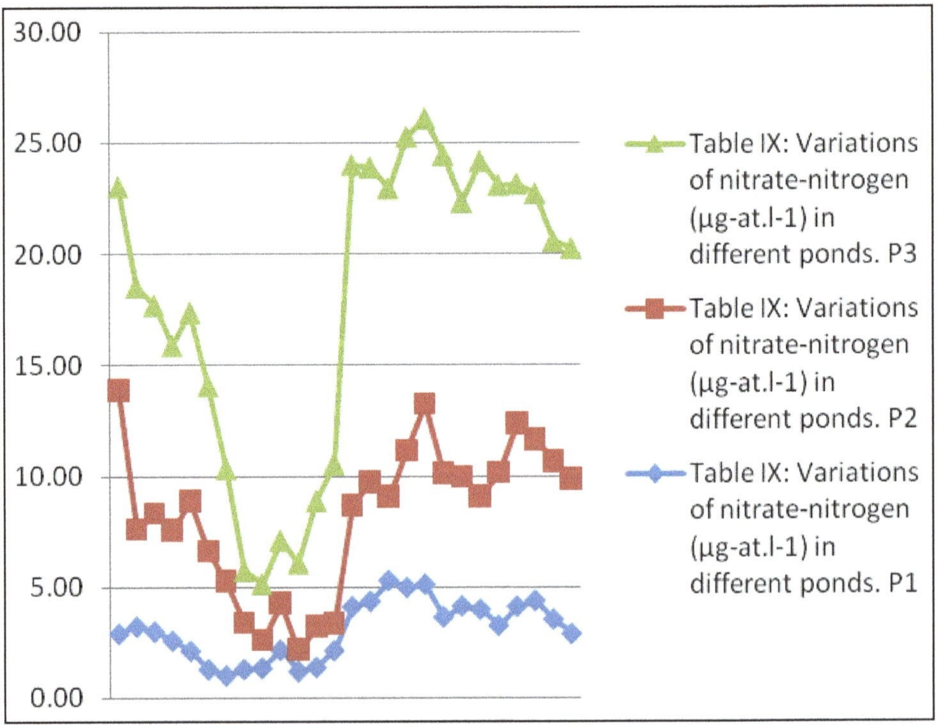

30.00
25.00
20.00
15.00
10.00
5.00
0.00
Table IX: Variations of nitrate-nitrogen (µg-at.l-1) in different ponds. P3
Table IX: Variations of nitrate-nitrogen (µg-at.l-1) in different ponds. P2
Table IX: Variations of nitrate-nitrogen (µg-at.l-1) in different ponds. P1

Figure 13.2: Variations of Nitrate-Nitrogen (µg-at.l^{-1}) in different Ponds.

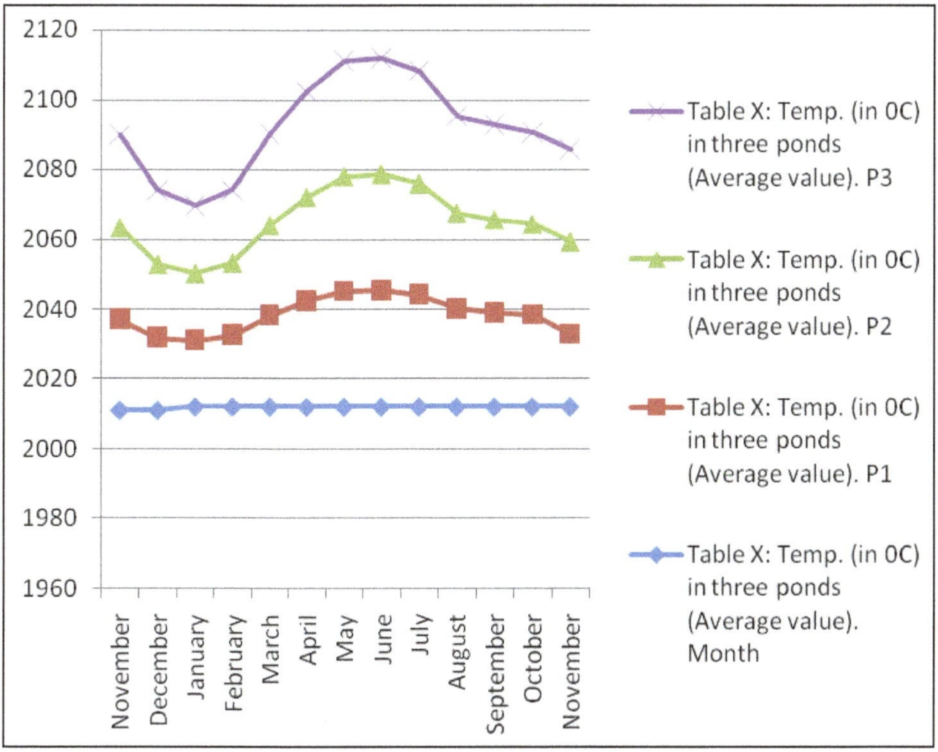

Figure 13.3: Temperature (in °C) in Three Ponds (Average values).

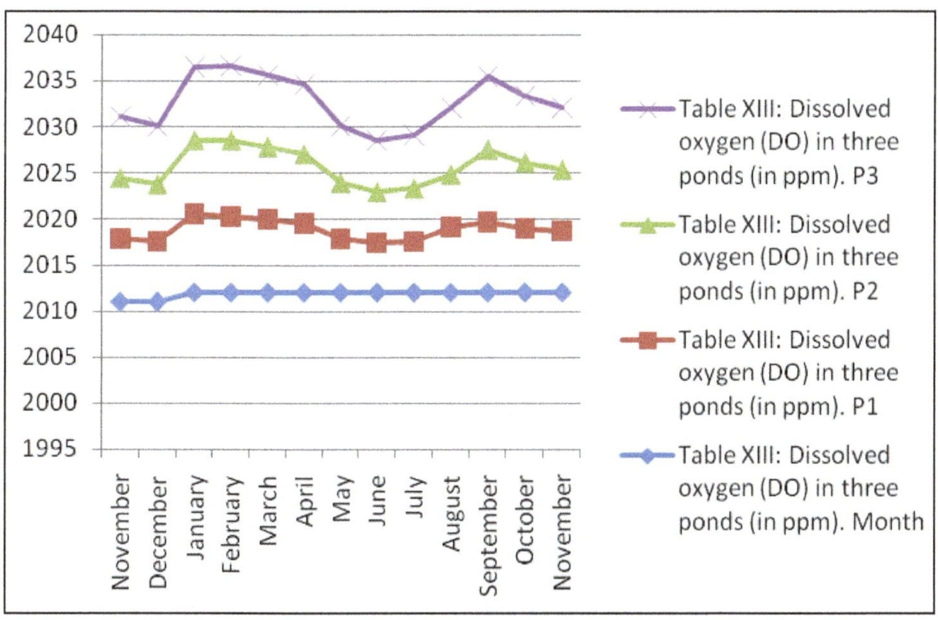

Figure 13.4: Dissolved Oxygen in Three Ponds (in ppm).

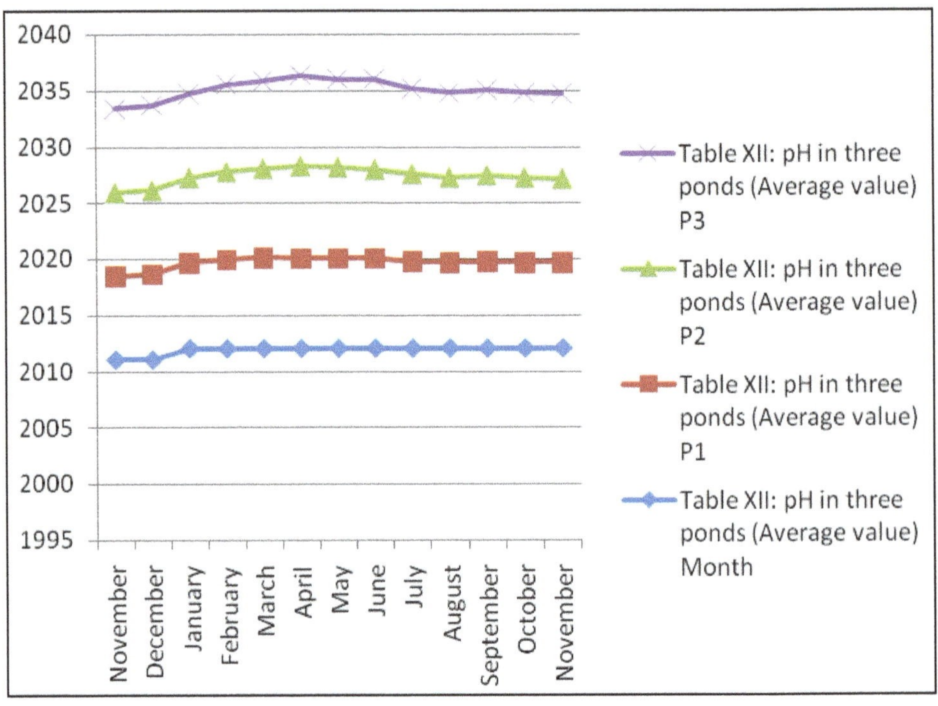

Figure 13.5: pH in Three Ponds (Average value).

Catla catla (surface feeder) *Labeo rohita* (column feeder) and *Cirrhina mrigala* (bottom feeder) and with different feeding habits occupying different zones of the ponds utilize the available food of the ponds in all the zones profitably. Regularly the weeds were removed by physical methods, manuring was done along with providing artificial diet like oil cakes, waste vegetables *etc.*, the fish were caught and weighed at intervals. The fish were caught by netting during March-April 2013. The weight of fish, ranged from 500gm-1000gm. Fishes below 500gm were again left in the ponds for further growth. The average yield of fishes and their cost at site was from P1-280kg, from P2-263kg and from P3-285kg @ Rs 100/kg and the total sale price was Rs. 82,800 in the year under study(2013).

Awareness programmes were organized periodically in the villages in order to create awareness among people in order to motivate them for fish cultivation. The resource persons were invited from Fishery Training Institute, Balugaon; WBUAFS, Kolkata; Berhampur University and from CIFA, Bhubaneswar. Training programmes on composite fish farming, were organized for Villagers and students. The technical expertise so gained is believed to be utilized by the beneficiaries for gainful earning. During the period of study care of the ponds was monitored by a group of peer volunteers from each village who have assisted in managerial activity and watch of the ponds in their respective villages. The profit of the sale proceeds of fish was being used as seed money by the volunteers for cultivation of fish for livelihood besides other engagements. Thus the objectives have been achieved through training and interaction sessions generating confidence among the villagers for aquaculture for their livelihood.

Species	P1 (3500Sq.ft)	P2 (3600 Sq.ft)	P3 (4000Sq.ft)
Catla	102 KG	118kg	130kg
Rohu	98kg	75kg	85kg
Mrigala	80kg	70kg	70kg
Total yield	280kg	263kg	285kg
Sale price	28,000	26,300	28,500
Approx. exp.	10,000	10,000	10,000
Profit (Rs)	18,000	16,300	18,500

REFERENCES

Adeniji, H.A. and Ovie, S.I. 1982. Study and appraisal for the water quality of the Ase Oli and Niger Rivers. *NIFFER Annual Report*, 15-20.

APHA. 1989. Standard methods for examination for water and wastewater, 17th edition. American Public Health Association. Washington DC.

Boyd, C.E. 1989. Water quality management and aeration in shrimp farming. Fishes and Allied Aquaculture Department Series. No. 2. Birmingham Ala Aubum University Press.

Boyd, C.E. 1990. Water quality in ponds for aquaculture. Albama agricultural experiment station, Auburn University Ala.

Das, S.K. and Padhi, S.N. 2014. In Application of Biology for Self Employment. Ed. Vol. Nanda Kishore Publication, Bhubaneswar.

EPA, 2006. Water Quality Standards Review and Revision, Washington DC.

Gupta, S.K. and Gupta, P.C. 2006. General and applied technology (Fish and Fisheries) S. Chand and Company, New Delhi.

Jhingran, V.G. 1985. Fish and fisheries of India. Hindustan Publishing corporation, Delhi, India.

Nikolosky, G.V. 1963. The ecology of fishes. Academic Press, London, U.K.

Olopade, daniyi 2013. Lakes, reservoirs and ponds. 7 (1): 9-19.

Padhi, S.N., Das, S.K.,Panda, A. and Panda, Sasmita. 2015. In Employment through aquaculture. Nanda Kishore Publication, Bhubaneswar.

Philips, M.J., Beveridge, M.C.M. and Clark R.M. 1991. Impact of aquaculture on water resourses. In D.E. Brune and J.R. Tomasso (eds), Advance in Aquaculture Vol.3: 568-591.

Zweig, R.D., Morton, J.D. and Stewart, M.M. 1999. Source water quality for aquaculture. A guide for assessment World Bank Report, 74.

Chapter 14

Plant Based Protein Sources: An Active Area of Research in Aquaculture Nutrition

☆ *Meenakshi Jindal and Rachna Gulati*

ABSTRACT

Fishmeal (FM) belongs to a short list of excellent feedstuffs that provide essential nutrients in a highly digestible concentrated form. The use of FM in domestic and farm animal diets will remain a core and efficient practice, particularly for young, rapidly growing, and high-producing animals like maturing fish, berried (egg-laden) shrimp, poultry, and lactating dairy cattle. The beneficial effects of eating wholesome foods will increase the worldwide demand for seafood products resulting in increased use of fishmeal. Optimal use of fishmeal in practical aquaculture diets is necessary to minimize feeding costs which can account for 40 per cent or more of operating expenses. The concentration of high-quality nutrients, especially protein, makes FM one of the most sought and expensive feedstuffs. The cost of high-quality FM (65 per cent protein) has ranged from 2.0 to 3.5 times the price of soybean meal. Therefore, fish diets can be made by partial or total replacement of FM with other plant and animal proteins. All-plant protein-based diets containing soybean meal, cottonseed meal, and middlings from corn and wheat, supplemented with lysine and methionine, have been used successfully to grow juvenile catfish, carp, and tilapia to market size. The aquaculture industry must continue to seek out alternative sources of high-quality plant based protein ingredients for their feedstuffs. Presently, this is an active area of research in aquaculture nutrition.

INTRODUCTION

The usual traditional diets *i.e.* rice bran and oil cakes do not contain enough nutrients required for good health of the fish. Further, these nutrients are not only low in energy but also yields a lot of wastes in the treated water. Therefore, to reduce pollution and increase fish growth, a nutritionally balanced diet containing fishmeal (FM) is necessary for enhancing not only the growth but also nutritionally rich fish flesh, such fishes also fetch better price in the market.

FM is a generic term for a nutrient-rich feed ingredient, sometimes used as a high-quality organic fertilizer. FM carries large quantities of energy per unit weight and is an excellent source of protein, lipids (oils), minerals, and vitamins; there is very little carbohydrate in FM.

FM is high quality animal protein. It contains Ca, P, K, Na and certain unidentified factors capable of promoting growth. 27 per cent of total world fish catch is converted to FM. Pelagic fish sp. such as Sardines, Mackerel and Anchovies support the world's FM industry. FM prepared from whole fish appears to be a better protein supplement than other animal protein.

Most commercial FM is made from small, bony and oily fish that otherwise are not suitable for human consumption and some is manufactured from by-products of seafood processing industries. FM can be made from almost any type of seafood but is generally manufactured from wild-caught, small marine fish that contain a high percentage of bones and oil. These fishes are considered 'industrial' since most of them are caught for the sole purpose of FM and fish oil production. The supply is presently stable at 6.0 to 6.5 million tons annually.

Approximately 4 to 5 tons of whole fish are required to produce 1 ton of dry fishmeal. The principal FM-producing countries are Chile, China, Thailand, U.S.A., Iceland, Norway, Denmark, and Japan (Table 14.1). Major groups of industrial fish rendered into FM are anchovies, herrings, menhaden, sardines, shads, and smelts (Table 14.2).

Table 14.1: Top Fishmeal Producing Countries

- ✰ **Peru** (Anchovy): It produces almost one-third of the total World FM supply.
- ✰ **Chile** (Anchovy and Horse mackerel).
- ✰ China (Various species).
- ✰ **Thailand** (Various species).
- ✰ **U.S.A.** (Menhaden, Pollock).
- ✰ **European Union, others** (Various species).
- ✰ **Iceland and Norway** (Capelin, Herrings, Bluewhiting).
- ✰ **Denmark** (Pout, Sandeel, Sprat).
- ✰ **Japan** (Sardine/Pilchard).
- ✰ South Africa (Pilchard).

Processing and Preservation of Fishmeal

Fish is a highly perishable raw material, and spoilage will occur if it is not processed in a timely manner. Preservation using ice or refrigerated seawater is common. Cooking, pressing, drying and grinding the fish make FM. There are several processing methods to produce good quality FM, but the basic principle

Table 14.2: Principal Fish Species in Fishmeal

Anchovies (Engraulidae):

- ☆ *e.g.,* Peruvian anchoveta (*Engraulis ringens*);
- ☆ Japanese anchovy (*Engraulis japonicus*).

Herrings, Menhaden, Sardines and Shads (Clupeidae):

- ☆ *e.g.*, Atlantic herring (*Clupea harengus*);
- ☆ Menhaden (*Brevoortia tyrannus* and *B. patronus*);
- ☆ South American and Japanese pilchards (*Sardinops sagax*) and other species; European pilchard (*Sardina pilchardus*); European sprat (*Sprattus sprattus*).

Smelts (Osmeridae):

- ☆ *e.g.,* Capelin (*Mallotus villosus*).

Jacks (Carangidae):

- ☆ *e.g.,* Chilean horse mackerel (*Trachurus murphyi*), Atlantic horse mackerel (*Trachurus trachurus*). Pollock, Cod, and Haddock (Gadidae) *e.g.*, Walleye or Alaska Pollock (*Theragra chalcogramma*);
- ☆ Atlantic and Pacific cods (*Gadus morhua* and *G. cephalus*);
- ☆ Georges Bank haddock (*Melanogramus aeglefinus*); Norway pout (*Trisopterus esmarkii*);
- ☆ Blue whiting (*Micromesistius poutassou*).

Hakes (Merlucciidae) and Sand lances (Ammodytidae):

- ☆ *e.g.,* Hake (*Merluccius* sp.);
- ☆ Hoki (*Macruronus novaezelandie*). Small and lesser sandeels (*Ammodytes marinus* and *Ammodytes tobianus*).

Tunas and Mackerels (Scombridae):

- ☆ *e.g.,* Skipjack tuna (*Katsuwonos pelamis*), Yellowfin tuna (*Thunnus albacares*);
- ☆ Chub mackerel (*Scomber japonicus*), Atlantic mackerel (*S. scombrus*).

Cutlassfishes (Trichiuridae):

- ☆ *e.g.* Largehead hairtail or Atlantic cutlassfish (*Trichiurus lepturus*).

Source: Miles and Chapman, 2006.

involves separation of the solids from the oil and water. When no oil needs to be removed, such as with lean fish, the pressing stage is often omitted. The general scheme of FM processing is given in Figure 14.1.

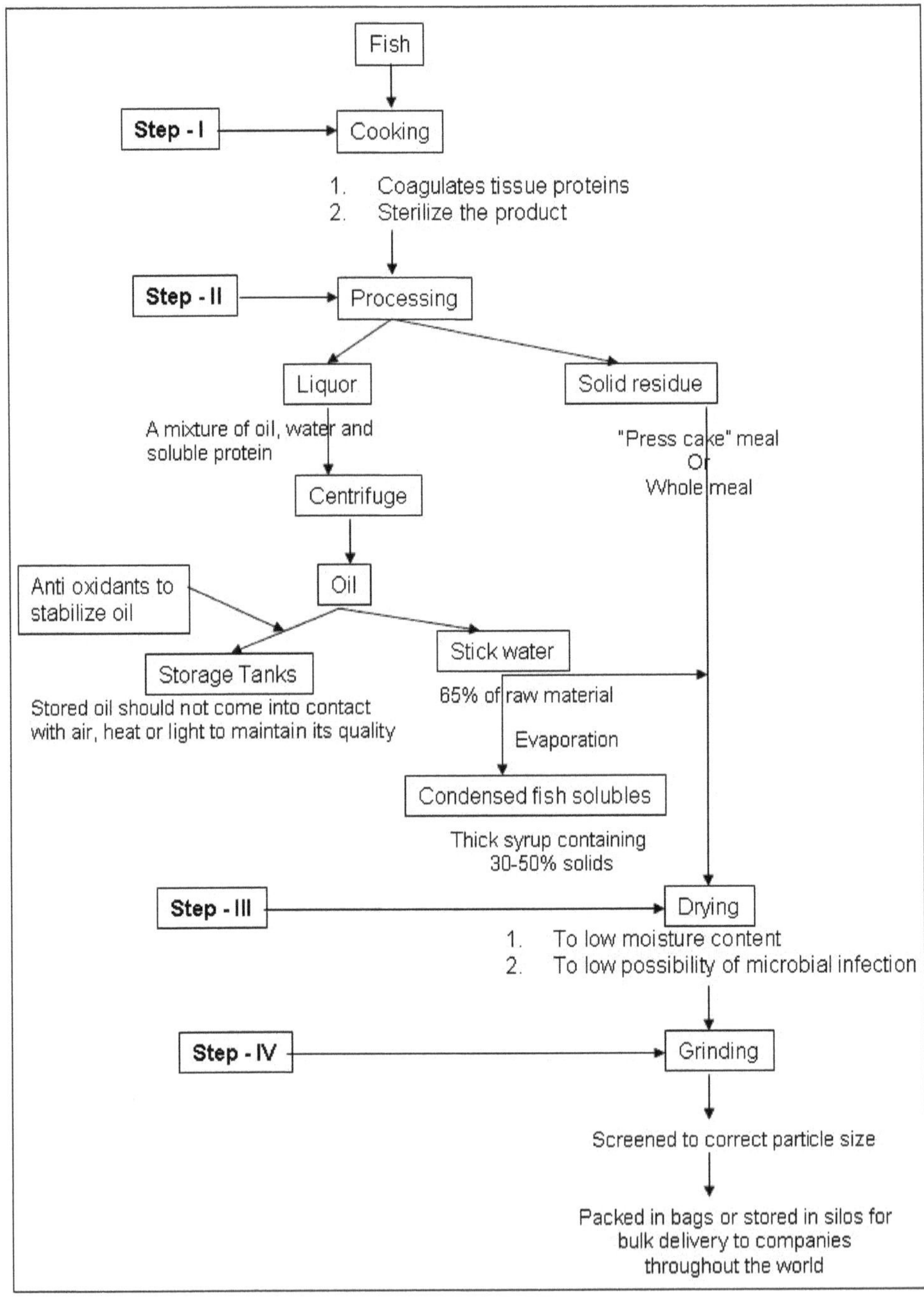

Figure 14.1: General Flowchart for Fishmeal Preparation.

The use of antioxidants in the preservation of FM is essential in order to stabilize its energy value because there are high quantities of polyunsaturated fatty acids (PUFAs) present in the oil. Without stabilizing FM with antioxidants the available energy content of the meal may be reduced by as much as 20 per cent; the oxygen will modify (damage) the chemical structure of the PUFAs and therefore less energy is available to the animal.

Protein Quality of Fishmeal

Any complete diet must contain some protein, but the nutritional value of the protein relates directly to its amino acid composition and digestibility (Table 14.3). Proteins are made of amino acids, which are released for absorption into the blood following protein digestion. Animals have requirements for specific amino acids rather than protein. FM and any other feedstuff that contains protein can simply be thought of as a 'vehicle' for providing amino acids to the diet.

Table 14.3: The Percentage Values for the Essential Amino-acid Composition of different Feedstuffs

Essential Amino Acid	*Fishmeal (64.5%)*[1]	*Meat and Bone Meal (55.6%)*[1]	*Poultry byproduct meal (59.7%)*[1]	*Blood Meal (89.2%)*[1]	*Soybean Meal (50.0%)*[1]
Arginine	3.82	3.60	4.06	3.75	3.67
Histidine	1.45	0.89	1.09	5.14	1.22
Isoleucine	2.66	1.64	2.30	0.97	2.14
Leucine	4.48	2.85	4.11	10.82	3.63
Lysine	4.72	2.93	3.06	7.45	3.08
Methionine + Cystine[2]	2.31	1.25	1.94	2.32	1.43
Phenylalanine + Tryosine[3]	4.35	2.99	3.97	8.47	4.20
Threonine	2.31	1.64	0.94	3.76	1.89
Tryptophan	0.57	0.34	0.46	1.04	0.69
Valine	2.77	2.52	2.86	7.48	2.55

1: Percentage of total crude protein in feedstuff.

2: Cystine can be synthesized from methionine.

3: Tyrosine can be synthesized from phenylalanine.

The values in parenthesis indicates the percentage of crude protein in the meal.

Source: NRC, 1993.

High-quality FM normally contains 60 to 72 per cent crude protein by weight (Table 14.4). From a nutritional standpoint, FM is the preferred animal protein supplement in the diets of farm animals and often the major source of protein in diets for fish and shrimp. But FM is very expensive and also deteriorates easily. Thus research is in progress to replace FM partially or wholly from fish feeds by using plant origin protein sources. A typical inclusion rate of FM in terrestrial livestock diets is usually 5 per cent or less on a dry matter basis.

Table 14.4: Gross Nutrient Value: Fishmeal vs Unprocessed Ingredients

	Ash	*N*	*Protein*	*Fat*	*GE*
Fishmeal	14	11	68	10	2
Blood meal	3	13	84	0	21
Meat and bone	33	8	53	7	16
Meat low ash	3	13	81	10	26
Poultry meal	14	9	57	17	21
Feather meal	3	12	74	10	22
Soybean	8	7	45	3	16
Lupin	3	5	32	5	17
Field pea	3	4	24	1	16
Cow pea	3	4	22	1	16
Wheat gluten	1	12	72	0	22
Corn gluten	1	9	58	0	22
Wheat	1	2	11	2	17
Sorghum	2	2	13	4	17

Lipid Content and Energy in Fishmeal

The lipids in FM provide a high content of energy to the diet. Since there is very little carbohydrate in FM, the energy content of FM relates directly to the percentage of protein and oil it contains. The quantity and quality of oil in FM will in turn depend on the species, physiology, sex, reproductive status, age, feeding habits of the captured fish, and the method of processing.

The lipids in FM and fish oil are easily digested by fish. The high digestibility of fish lipids means they can provide lots of usable energy (Table 14.4). If a diet does not provide enough energy, the fish or shrimp will have to break down valuable protein for energy, which is expensive and can increase production of toxic ammonia.

Fish lipids are excellent sources of the PUFAs in both the omega-3 and omega-6 families of fatty acids. The predominant omega-3 fatty acids in fishmeal and fish oil are linolenic acid, DHA and EPA. PUFAs can be easily damaged and become rancid when exposed to oxygen, a process known as oxidation and one that releases heat.

Mineral Value of Fishmeal

When a sample of feed is taken to the laboratory and analyzed for nutrient content, the procedure involves burning a portion of the sample. Ash is the material remaining after the feed sample is completely burned. Normally, the ash content of good quality averages between 10-17 per cent. More ash indicates a higher mineral content, especially calcium, phosphorus, and magnesium. Calcium and phosphorus constitute the majority of the ash found in FM.

Vitamin Value of Fishmeal

The vitamin content of FM is highly variable and influenced by several factors, such as origin and composition of the FM processing method, and product freshness. The content of fat-soluble vitamins in FM is relatively low because of their removal during extraction of the oil. FM is considered to be a moderately rich source of vitamins of the B-complex especially cobalamine (B12), niacin, choline, pantothenic acid, and riboflavin.

Reduction in Global FM Production

The high quality and concentration of essential nutrients, especially of well-balanced amino acids, essential fatty acids, and energy content makes FM an indispensable ingredient in diets of most aquaculture species and many land-farm animals. Because of its nutrient content, high digestibility and palatability, FM serves as the benchmark ingredient in aquaculture diets. FM belongs to a short list of excellent feedstuffs that provide essential nutrients in a highly digestible concentrated form. The use of FM in domestic and farm animal diets will remain a core and efficient practice, particularly for young, rapidly growing, and high-producing animals like maturing fish, berried (egg-laden) shrimp, poultry, and lactating dairy cattle. The beneficial effects of eating wholesome foods will increase the worldwide demand for seafood products resulting in increased use of fishmeal.

Global FM production has been relatively constant in the past 15 years, but production declines (Figure 14.2) as much as 15 per cent when *El Nino* effects are

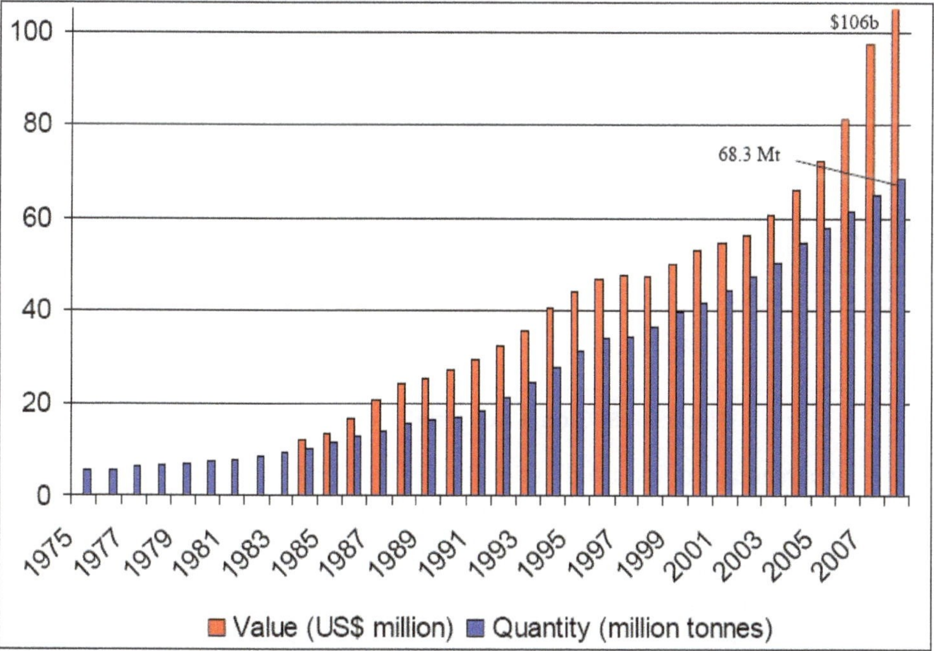

Figure 14.2: Global Aquaculture (million tones and US$ million); FAO Fishstat 2009.

severe (FAO FishStat, 2009). EI Nino causes Ocean warming off the coasts of Peru and Northern Chile, causing anchovies to move elsewhere, thus reducing the availability of anchovies that support's the world's largest fishery for FM production. Although Peru and Chile account for less than one-third of Global FM production, they produce a large proportion of FM traded throughout the World, more than 60 per cent. Other large FM producing countries, such as Norway, utilize nearly all of their production domestically.

The issue of limited global supplies of FM and fish oil is becoming critical. The per cent age of global FM production used in fish feeds has increased from less than 10 per cent prior to 1990, to more than 50 per cent in 2007 (Figure 14.3). In the case of fish oil, the situation is even more critical. Prior to 1990, less than 10 per cent of global fish oil production was consumed annually by the fish feed manufacturing sector, but by 2007, the per cent age increased to more than 80 per cent, mainly at the expense of use of fish oil to produce margarine (Figure 14.4). If current production trend continue upward, fish oil supplies will be inadequate to meet the needs of fish feed manufacturing industry within a few years.

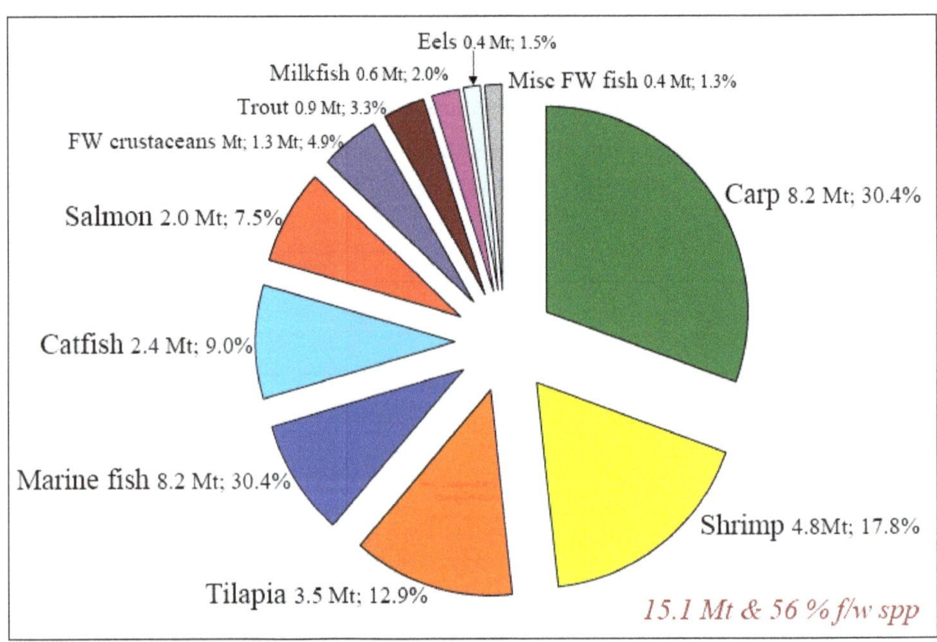

Figure 14.3: Global Aquafeed Production in 2007.

The percentage of FM used in fish feed is predicted to decrease by 10 per cent and 5 per cent respectively, by 2010 (Figure 14.4). This change will require higher use levels of alternative protein sources, including rendered products (where allowed) and oilseed proteins, mainly soybean meal, to replace the FM protein (Figure 14.5). Even greater reduction in fish oil use will be needed, perhaps as much as 50 per cent of current use levels. Plant oils will be needed to replace the fish oil.

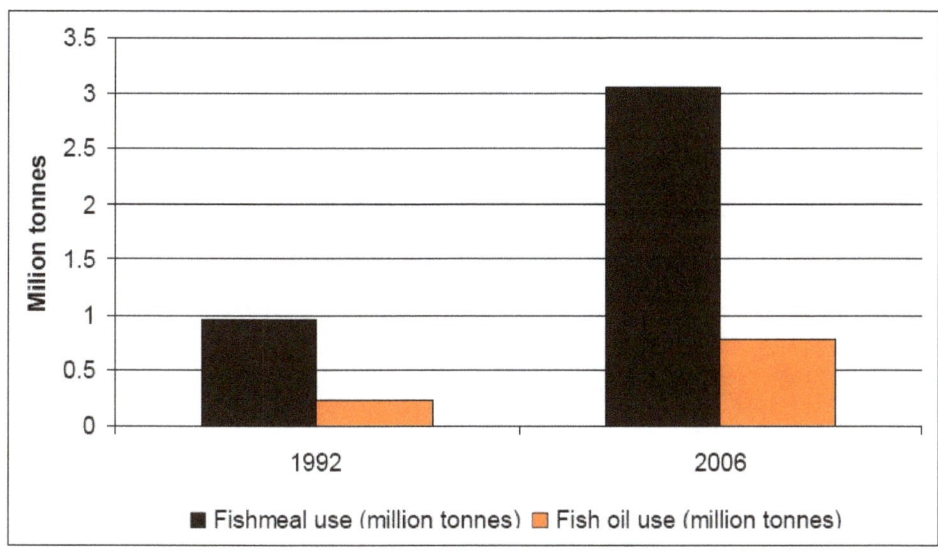

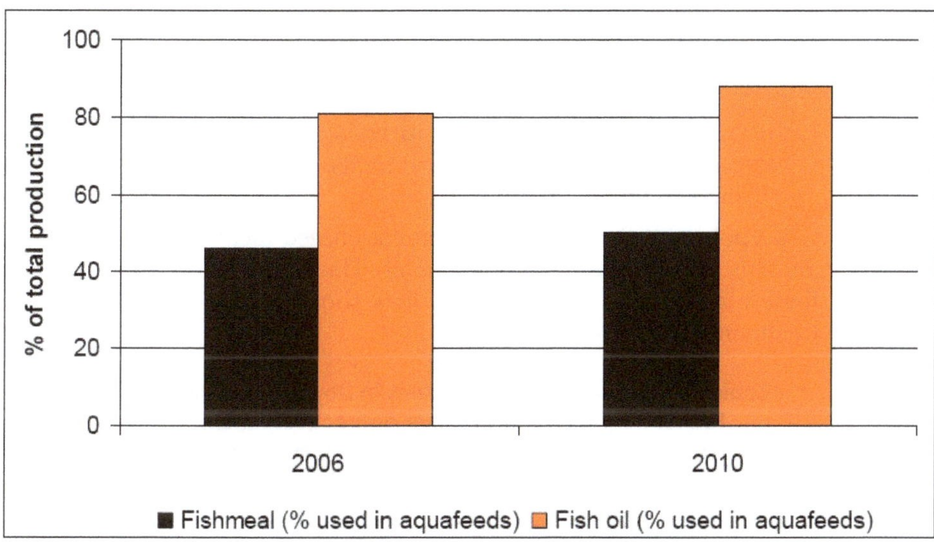

Figure 14.4: Fishmeal and Fish Oil Use in Aquafeeds.
***Source*: http://seafish.sin.org.**

Proteins isolated from plants are associated with indigestible non-structural carbohydrates (oligosaccharides) and structural fiber components (cellulose), which are not associated with animal proteins. It is the presence of these components which are thought to be contributing obstacles to efficient utilization of proteins in many economically plant-based feedstuffs (Garg, 2003). Some examples showing presence of ANF's in plant protein sources are:

1. A naturally occurring anti-nutritional factors (ANFs) in uncooked soybeans is the Kunitz trypsin-inhibitor that prevents the enzyme trypsin from breaking down dietary proteins in the intestine of animals.

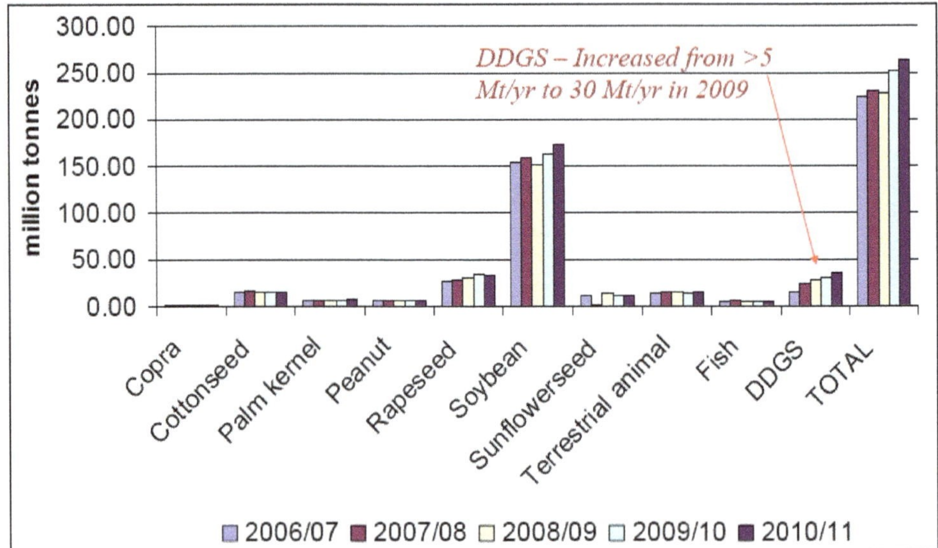

Figure 14.5: Global Production of other Protein Meals for Aquafeeds. *Source*: USDA Oilseeds: World Markets and Trade; DDGS – Ethanol Producer Mag. 2/9/2010.

2. Lathyrogens in chickpeas also disrupt collagen formation. Collagen is the most abundant protein present in animals, making up most connective tissue and providing structural support.
3. Gossypol is another ANF found in cottonseed meal/oil that is toxic to animals and lowers fertility in males. Therefore, before incorporating these plant proteins in fish feeds, they should be treated with suitable treatments (Table 14.5).

Table 14.5: Anti-nutritional Factors in Soy Products and Processing Steps to Remove or Inactivate them

Anti-nutrient	*Heat Sensitive*	*Extractable*	*Other Treatment*
Trypsin inhibitors	Yes	No	No
Hemagglutinin	Yes	No	No
Phytic acid	No	No	Phytase
Saponin	No	Yes	No
Phytoestrogen	No	Yes	No
Anti-vitamin	Yes	?	No

Replacement of FM with other Plant and Animal Proteins

In the past decade, the per cent age of FM used in fish feeds has decreased by about 50 per cent as a result of the use of alternative protein ingredients. Soybean meal was first tested in dry feed mixtures in the early 1940's by the Cortland research team and the feed cost per unit of production was reduced by about half (Tunison *et al.*, 1941). Addition of a vitamin mixture to the dry pellet formulation permitted the successful rearing of trout to spawning and subsequent rearing of fry (Phillips *et al.*, 1964).

Tilapia, carp, milkfish (SBM 20-60%, CGM 5-10%, R/CM 20-40%, CSM 1-25%, SO 1-8%),

Salmon & trout (SBM 3-12%, WGM 2-10%, CGM 10-40%, R/CM 3-10%, LKM 5-15%, FBM 5%, FPM 3%, R/CO 5-15%, SO 5-10%),

Grey mullet (SBM 20-25%), F/W prawns (SBM 15-25%),

Shrimp (SBM 5-40%, WGM 2-10%, CGM 2-4%, R/CM 3-20%, LKM 5-15%),

Marine fishes (SBM 10-25%, SO 3-6%, WGM 2-13%, CGM 4-18%, R/CM 7-20%, CPC 10-15%),

SBM – soybean meal; CGM – corn gluten meal; CSM – cottonseed meal; R/CM – rapeseed/canola meal; G/PM – groundnut/peanut meal; MC – mustard seed cake; LKM – lupin kernel meal; WGM – wheat gluten meal; FBM – faba bean meal; SO – soy oil; R/CO – rapeseed/canola oil

Figure 14.6: Results for Soybean and other Vegetable Ingredients in Aquafeeds.

New information on nutrient requirements of aquatic organisms coupled with advances in feed technology indicates that species-specific fish diets can be made by partial or total replacement of FM with other plant proteins (Jindal, 2002; Jindal and Garg, 2005). All-plant protein-based diets containing soybean meal supplemented with lysine and methionine, have been used successfully to grow juvenile catfish, carp, and other fish species to market (Jindal *et al.*, 2008, 2009). Animal proteins and fats can be used in aquaculture diets because they also provide essential amino acids and fatty acids. These "fishmeal substitutes" will be used more extensively by the aquaculture industry in the future.

Environmental Considerations in Fish Farming

Nitrogen (N) and phosphorus (P) are essential elements for living organisms and exist in water bodies as dissolved and particulate forms. N is essential for the normal growth of the fish and is an important ingredient in fish feed but is very expensive. P is required for the optimum growth, feed efficiency, bone development and maintenance of acid-base regulation in fish.

The aim of aquaculture should be to provide sufficient N for good growth through balanced feed. N pollution from aquaculture can occur in three ways, *viz.* (*1*) overfeeding of fish at a time when they are not growing (2) feeding unstable and highly soluble diets (3) providing a diet of poor absorption and nitrogen retention efficiency.

The main product excreted by teleost fish is total ammonia nitrogen, which is formed in lever and excreted across the gills. About 80-90 per cent of N loss from fish is through gill excretion and the faecal N loss accounts for 10-20. N is also lost through uneaten feed or dust.

P in hatchery effluents is present in two forms: solid particles (*e.g.* bones and other insoluble forms) and soluble P excreted by fish in urine. Although particulate P can be collected and removed but soluble P can not be removed economically because it is present in very low concentration in very large quantities of water. Thus, limiting the amount of digestible P in feeds to the amount needed by the fish is the approach used to produce low-polluting feeds. Using this approach, the amount of soluble P excreted by fish has been reduced to very low levels.

High ash feed ingredients, such as FM, meat and bone meal and poultry byproduct meal contain high amounts of N and P (Sugiura *et al.*, 2000). Plant protein ingredients such as soybean meal, are low P ingredients. Replacing FM with soybean meal lowers N and P content of the diet (Vielman *et al.*, 2000, Jindal, 2011 and 2013). However, about 75 per cent of P found in soybean meal is bound to phytic acid. P can be released from phytate by the enzyme phytase, a natural constituent of seeds, but its activity is destroyed by the heat associated with pelleting.

Comparison of Fishmeal with Plant Protein Sources

1. Amino-acid Profile

The amino acid profile of FM is what makes this feed ingredient so attractive as a protein supplement (Table 14.3). Proteins in cereal grains and other plant concentrates do not contain complete amino acid profiles and usually are deficient in the essential amino acids lysine and methionine. Soybean and other legume meals, which are widely used in the diets of most farm animals such as pigs and chickens, are a good source of lysine and tryptophan but are limiting in the sulfur-containing amino acids methionine and cystine. An animal's requirement for a limiting amino acid can be met by simply adding more of the protein. However, this would be very costly.

2. Protein Digestibility

Plant-based proteins, even when properly processed, are usually not as digestible as FM; and their inclusion rate into the diet is often limited as it results in depressed growth rates and feed intake. Over-all protein digestibility values for FM are consistently above 95 per cent. In comparison protein digestibility for many plant-based proteins varies greatly.

3. Source of Omega-3 and Omega-6 fatty Acids

FM and oil contain more omega-3, than omega-6 fatty acids. In contrast, most plant lipids contain higher concentrations of omega-6 fatty acids. For example, oil extracted from soybeans, corn, or cottonseed is rich in linoleic acid, an omega-6 fatty acid. Some oils, like those from canola and flax seeds contain linolenic acid (of the omega-3 family), however, its conversion into essential DHA and EPA by most animals may be limited.

4. Palatability

FM contains certain compounds that make the feed more acceptable and agreeable to the taste (palatable). This property allows the feed to be ingested rapidly, and will reduce nutrient leaching. It is thought the non-essential amino acid glutamic acid is one of the compounds that imparts palatability to FM.

5. Nutritional Inhibitors or ANF's

ANF's are compounds that interfere with nutrient digestion, uptake, or metabolism and can also be toxic. The structural nature of plants is totally different from that of animals. Proteins isolated from plants are associated with indigestible non-structural carbohydrates (oligosaccharides) and structural fiber components (cellulose), which are not associated with animal proteins. Therefore, plant proteins should be properly processed to remove ANFs before their incorporation in fish feed (Table 14.5). The lack of ANF's in FM also makes this meal more attractive than plant proteins for use in aquaculture diets.

Current Use Levels of Soybean as Protein

Estimates of future protein requirements for aquaculture feeds depends on future production from various segments of the aquaculture industry and the supply of FM. Results for soybean and other vegetable ingredients are given in Figure 14.56

Principle Objectives in the Formulation of Fish Feeds

1. Balancing nutrients in diets by using the minimum amount of FM to meet specific amino acid requirements for fast growth and reproduction and reducing feed costs constitute one of the principal objectives in formulation of fish feeds.
2. Another important aim in feed formulation is to increase dietary nutrient density and digestibility of the feed to increase biological performance and to reduce nutrient leaching and water-quality degradation.

Future Outlook and Conclusions

Aquaculture is the fastest growing segment of livestock production worldwide. It is expected to exceed cattle farming in terms of total production by 2020. Aquaculture is growing in large part because production is switching to more intensive, higher-input systems. The main input is feed and feed production for aquatic animals is expected to nearly triple by the end of the decade. This growth absolutely dictates greater use of plant protein sources other than FM.

Fish fed diets formulated with a high percentage of FM will contain high concentrations of the PUFAs in their tissues, especially of the omega-3 family. PUFAs are essential for human biological functions, particularly the production of prostaglandins. PUFAs and prostaglandins may ameliorate many human health disorders, such as high blood pressure, heart disease, arthritis, migraines, diabetes, and cancer. Incorporation of DHA and EPA found in FM into the diets of fish and other farm animals is an efficient method to ensure a proper concentration of these important omega-3 fatty acids in the human diet.

The aquaculture industry must continue to seek out alternative sources of high-quality plant protein ingredients as fish feedstuffs. Soy products are the leading candidates to supply this protein. Presently, this is an active area of research in aquaculture nutrition.

REFERENCES

Garg, S. K. 2003.Recent methods in fish nutrition and biotechnology.Practical Manual, Department of Zoology and Aquaculture, CCS HAU, Hisar pp.1-183.

Jindal, Meenakshi 2001. Effect of feeding of plant origin proteins on nutrient utilization and growth of *Channa punctatus* (Bloch.).Ph.D. Thesis. CCS HAU, Hisar pp. 141.

Jindal, Meenakshi and S. K. Garg. 2005. Effect of replacement of fishmeal with defatted canola on growth performance and nutrient retention in the fingerlings of *Channa punctatus (bloch.)Pb. Univ. Res. J. (Sci.)*, 55: 183-189.

Jindal, Meenakshi, Garg, S.K. and Yadava, N.K. 2008. Effect of replacement of fishmeal with processed soybean on daily excretion of ammonical - nitrogen (NH_4-N) and ortho-phosphate (O-PO_4), in *Channa punctatus* (Bloch.). *Panjab University Research Journal (Science)*,58: 25-33.

Jindal, M., Garg, S.K. and Yadava, N.K. 2009. Effect of feeding defatted canola on daily excretion of ammonical nitrogen (NH_4-N) and ortho-phosphate (o-PO_4) in *Channa punctatus* (Bloch). *Livestock Research for Rural Development, USA. Vol. 21, Article # 35* Retrieved from http: / /www.Irrd.org./lrrd21/3/jind21035.htm

Jindal, Meenakshi. 2013. Effect of Diel Rhythms of Feeding on the Growth Performance of the Walking Catfish, *Clarias batrachus* Fingerlings. *Fishery Technology*, 50: 11-15.

Miles, R.D. and Chapman, F.A. 2006. The benefits of fishmeal in Aquaculture diets. In: Catfish Protein Nutrition (Ed. Robinson, Edwin H. and Menghe, H. Li.) pp 51-59.

NRC (National Research Council), 1993. Nutrient requirements of fish. Natiuonal Academy of Science, Washinton, D.C., 102p.

Phillips Jr., A. M., Podoliak, H. A., Livingston, D. L., Booke, H. E. and Pyle, E. A. 1964. The Nutrition of trout. *Fisheries Research Bulletin*, 27: 47-56.

Sugiura, S. H., Babbitt, J. K., Dong, F. M. and Hardy, R. W. 2000. Utilization of fish and animal byproduct meals in low pollution feeds for rainbow trout, *Oncorhynchus mykiss* (Walbaum). *Aquaculture Research*, 31 (7), 585-593.

Tunison, A. V., Phillips, A. M. and Brockway, D. R. 1941. The nutrition of Trout. *Fisheries Research Bulletin*, 1 (10): 3-20.

Vielman, J; Makinen, T.; Ekholm, P. and Koskela, J. 2000. Influence of dietary soybean and phytase levels on performance and body composition of large rainbow trout (*Oncorhynchus mykiss*) and algal availability of phosphorus load. *Aquaculture* 183: 349-362.

Chapter 15

Reservoir Fisheries of India

☆ *V.B. Sakhare*

Reservoirs are artificial lakes which are formed by impounding rivers or streams by way of construction of dams across them. The reservoirs are mainly constructed for irrigation, hydel-power generation and flood control. The development of fishery in reservoir is coming as the last. China, the world leader in reservoir fisheries records a mean yield of 7800 kg/ha from its reservoirs and those from Sri Lanka (7300 kg/ha) and Cuba (7100 kg/ha) are also high yielding. The average productivity from all categories of Indian reservoirs is estimated at about 15 kg/ha/yr, although potential exists for manifold increase as demonstrated in small, medium and large reservoirs in the country. Nearly one million tonnes of fish can be produced from reservoirs every year by adopting fisheries management norms based on scientific advice (Dixitulu, 2010).

The reservoir fisheries increase self employment opportunities in the rural areas. The success stories of scientifically managed reservoirs in from different states of our country testified that with scientific management fish production from all categories of reservoirs can be increased substantially.

Natarajan (1974) estimated that the average annual fish yield from Indian reservoirs was capable of increasing to 75 kg/ha which, however,was toned down later to 60 kg/ha in 1981.Sreenivasan *et.al*; (1980) estimated the production potential of Indian reservoirs at 80 kg/ha/yr and 200 kg/ha/yr. Even according to a conservative administrative estimate (Yadava, 1982), the potential yield of Indian reservoirs is around 50 kg/ha/yr by 2000 A.D. According to Indian institute of management, Ahemadabad (1985), the average annual fish yield is estimated to increase to about 60 kg/ha.

Estimates regarding actual yield per hector of Indian reservoirs varied from 5-8 kg/ha by National Commission on Agriculture in 1976, 10 kg/ha by Jhingran (1983) and 14.5 kg/ha by IIM, Ahemadabad (1985). There is a declining trend in the per hectare yield of certain important reservoirs like Rihand, Ukai, Keetham *etc.* (Saxena, 1990) and the yield for hectare in most of the small reservoirs is not appreciably higher than the medium reservoirs. Even the overall position of AICRPRF (All India Coordinated Research Project on Reservoir Fisheries) reservoirs is in no way better than other reservoirs (Saxena, 1990). Thus, the Indian reservoirs not only produce much less than their potential, their yield rates are appreciably lower that their Asian counter parts.

There are various estimates on the total area under Indian reservoirs. The National Commission on Agriculture (1976) has estimated the total area under reservoirs at 3 million ha during the mid-sixties and projected its growth to 6 million ha by 2000 A.D. Sugunan (1995) estimated a total of 19370 reservoirs in country with a total area of 3.15 million ha. However, the fisheries statistics of the Ministry of Agriculture (Dept. of Animal Husbandry and dairying) (Aran, 1999) gives an estimate of 2.03 million ha under reservoir fisheries.

Among the different states, Tamil Nadu has the highest number of small reservoirs, followed by Karnataka and Andhra Pradesh. So far as medium category of reservoirs are concerned, Madhya Pradesh is at the top in total area as well as in water spread area, followed by the states of Karnataka and Andhra Pradesh. The maximum number of large reservoirs is in the state of Karnataka but the total area is less when compared to that of Andhra Pradesh. Considering the vastness of the resources and its production potential, these reservoirs have become the main inland capture fisheries resources, for the future fisheries development of India.

Classification of Reservoirs

Regarding classification of reservoirs, there was an anomaly which existed for a long time, but now Indian reservoirs are classified on the basis of water spread area. Based on the water spread area Indian reservoirs are classified as small (above 10 ha to <100 ha), medium (1000 ha to 5000 ha), and large (>5000 ha). Small reservoirs occupy 1.49 million ha followed by large (1.14 million ha) and medium (0.52 million ha).

Among variously sized reservoirs, maximum annual production is reported from small reservoirs (50 kg/ha), followed by medium (12.5 kg/ha) and large (11.4 kg/ha) with an average of 20.13 kg/ha. Despite the amenability for fish production and a production potential in the range of 50-300 kg/ha, the present yield from Indian reservoirs is very low (Vass, 2006).

Management of Small Reservoirs

Separate management norms are followed for large and small reservoirs. Small reservoirs offer the easiest way to obtain much enhanced fish production at minimum cost. Nevertheless, the investigations conducted by CIFRI over the past few decades have resulted in many guidelines, based on which the reservoir fishery managers can develop location-specific management norms. Such guidelines are more effective for the small reservoirs, where the relationship between management and yield improvement is known to be more precise compared to large impoundments.

Development of fisheries of small reservoirs is undertaken mostly by the application of extensive aquaculture practices based on the status of achieving culture-capture balance.Such impoundments are developed through large scale stocking and harvesting after allowing a grow out period of about 9 months. These shallow reservoirs which get mostly dried up during summer, allow complete harvesting of fish mostly leaving no broodstock for a succeeding fishery through natural recruitment. Therefore, by and large, the fisheries have to be built up year after year through regular stocking. Stocking policy is to be evaluated on biogenic capacity of reservoir concerned giving allowance after escapement of broodfishes and youngones through spillways and irrigation canals. Absence of predators in such water bodies permits early stocking of advanced fry instead of fingerlings, so as to get maximum grow out period (Desai, 2008).

Application of stock enhancement in specific small reservoirs has shown the feasibility of this approach in significantly raising the fish yield from such water bodies (Table 15.1).

Small reservoirs are shallower and biologically more productive per unit area than larger reservoirs. Stocking with fast growing fishes has proved to be the most useful tool in developing fisheries of small reservoirs. This has proved to be highly remunerative where almost complete annual harvesting is possible. Such systems have therefore been categorized as 'put and take' systems. However, stocking needs to be carefully planned based on the biogenic capacity of the reservoir, growth rate of the stocked species, natural mortality, escapement through the irrigation canal and spillway, and the predator pressure, if any seed stocking can even be maintained at a higher level than the natural carrying capacity of the environment through supplementary fertilization (Jhingran, 1990).

Management of Large and Medium Reservoirs

The large and medium reservoirs are predominantly capture systems and their management norms are based on the principle of stock manipulation, by adjusting fishing efforts, observance of closed season and gear selectivity.

Since large reservoirs, in the long run, are to be developed into capture fisheries, it is imperative that the stocked fish breed in the reservoir resulting in autostocking.

In large reservoirs, the management measures centre round the development of optimum fishing effort and selection of right types of tackle. Appropriate measures to conserve the desirable species and maintenance of the right balance in species mix is also ensured through stocking, mesh regulations, protection of breeders, *etc.*

In Govindsagar reservoir, the three-pronged strategy of enlarged mesh size, increase in fishing effort and stocking has paid rich dividends. In 1973 the mesh size was enlarged from 65 mm to 100-180mm bar and the fishing effort was increased by 120 per cent.There was sustained stocking of common carp which responded well to the reservoir environment. All these steps helped to push up the production by about 200 per cent in just five years (Jhingran, 1988).

Factors Affecting Reservoir Productivity

Morphometry is an important factor in productivity of reservoirs. Rawson (1951, 1955) proved that mean depth is a dominant factor in productivity-which is inversely proportional to mean depth. Temperature is another dominant factor in productivity.

Based on the morphometric and edaphic factors Ryder (1965) evolved the concept of morphoedaphic index as tool for predicting fish yields.

MEI=TDS/Z (*i.e.*, total dissolved solids/Mean depth).

However in Indian reservoirs, which have wide annual water level fluctuations, heavy inflow-outflow, different species mix and changes in fish populations *etc.* this principle is not strictly applicable (Sreenivasan,1989).

Siltation

When river is tilled behind a dam, the sediments it contains sink to the bottom of the reservoir. The proportion of a river's total sediment load captured by a dam-known as its 'trap efficiency'-approaches 100 per cent for many reservoirs, especially in large reservoirs. As the sediments accumulate in the reservoir, so the dam gradually loses its ability to store water for the purposes for which it was built. Every large reservoir loses storage to sedimentation, although the rate at which this happens varies widely. Sedimentation is still the most serious technical problem faced by the large reservoirs. The rate of reservoir sedimentation depends mainly on the size of a reservoir relative to the amount of sediment flowing into it. Large reservoirs in the United Sates lose storage capacity at an average rate of around 0.2 per cent per year, with regional varieties ranging from 0.5 per cent per year in the Pacific states to just 0.1 per cent in reservoirs in the northeast. Major reservoirs in China lose capacity at annual rate of 2.3 per cent (Gleick, 1995).

Excessive siltation leading to drastic reduction in the water holding capacity is a common problem in Indian reservoirs. Siltation hampers the productivity of reservoir by affecting the life process of biotic communities. The catchment areas, especially those of the Ganga river are characterized by a prolonged dry season followed by a turbulent monsoon, with river discharges up to 8500 m^3S^{-1}.Thus,heavy erosion and high sediment load are characteristic of Indian rivers. The Indian rivers carry about 2050 million tonnes of silt, of which nearly 480 million tones is deposited in the reservoirs and 1572 tonnes is washed away into the seas. Loss of storage capacity of the reservoirs due to siltation is one of the most serious consequences of soil erosion. Many of the Indian reservoirs recorded siltation rates much excess of what was envisaged during the planning stage of the project (Sugunan, 1995).

Seed Stocking

The enhancement of reservoir fisheries through stocking of fish seed has been a common practice throughout the world. It has emerged as one of the most widespread management tools for reservoir fisheries, for the reason that it has often been biologically successful.

Stocking density is an important factor that determines the success or failure of the whole programme and hence this needs to be assessed carefully. To determine the optimum stocking density, factors such as ecological characteristics, and existing biomass should also be considered.

The optimum size for stocking is mainly chosen by considering twp factors; viz; cost and survival. Stocking of fish at very small sizes leas to risks associated with high mortality even though the cost of stocking material increases exponentially with length. This mortality size relationship also provides the basis for a quantitative assessment of stocking density (Lalrinsanga *et al.*, 2006).

A number of formulae are available for computing the stocking density. A simple method,more easily adaptable in small reservoir is:

$$\text{Rate of stocking} = \frac{\text{Production in kg}}{\text{Individual growth rate in kg}} + \text{Loss}(\%)$$

Scientific literature is also replete with methods for calculating the potential from the reservoirs. The most popular among them is the morpho-edaphic index method, which is calculated as:

$$\text{MEI} = \frac{\text{TDS}}{\text{d}}$$

where,

MEI = Morpho-edaphic Index

TDS = Total Dissolved Solids

d = Mean depth of reservoir

Potential fish yield is calculated as:

Y = KXa

where,

Y= Potential yield

X= MEI

K = A constant representing the climatic effects 'a' is the exponent approximately 0.5

Selection of Species for Stocking

Selection of fishes for stocking is based on the assessment of existing biotic communities and their efficiency in converting primary trophic resources to harvestable products. Reservoir fisheries in India largely centers on development of carp fisheries. Major carps, by virtue of their feeding habits close to primary producers and their fast growth rate, are indispensable in reservoir fisheries management. The introduction of exotic fishes in Indian reservoirs is still a subject of controversy due to possible deleterious effects on indigenous populations.

In India, stocking of reservoirs mostly consists of fast growing compatible species of Indian major carps (*Catla catla, Labeo rohita* and *Cirrhinus mrigala*) for utilization of various food niches. Stocking rates need to be fixed for individual water bodies. Advanced fingerlings of carp (10-15 cm) may be stocked at 250 nos/ha in large reservoirs free from catfishes and 500 nos./ha in catfish dominated reservoirs (Desai,2008).stocking of fingerlings (4"-6" size) in small reservoirs has proved to be a useful tool for developing their fisheries (Sinha,1998).

Fish Production

Depending on the available literature, the fish production from small, medium and large sized reservoirs in India varies from 0.52 to 921.7 kg/ha/yr (Sakhare, 2007).The highest rate of fish production is recorded from small reservoirs and lowest from large sized reservoirs. A list of few scientifically managed reservoirs is depicted in Table 15.1.

Table 15.1: Production Trends in Scientifically Managed Reservoirs

Reservoir	*Category*	*State*	*Fish Production*	*Reference*
Aliyar	Small	Tamil Nadu	202 kg/ha/yr	CIFRI
Tirumoorthy	Small	Tamil Nadu	134 kg/ha/yr	CIFRI
Amaravathy	Small	Tamil Nadu	500 kg/ha/yr	Desai,2008
Gularia	Small	Uttar Pradesh	100 kg/ha/yr	CIFRI
Bachhra	Small	Uttar Pradesh	139 kg/ha/yr	CIFRI
Tungabharda	Large	Karnataka	91 kg/ha/yr	Desai,2008
Gobindsagar	Large	Himchal Pradesh	135 kg/ha/yr	Desai,2008
Pong	Large	Himchal Pradesh	53 kg/ha/yr	Desai,2008
Vallabsagar (Ukai)	Large	Gujarat	160 kg/ha/yr	Desai,2008
Adan	Small	Maharashtra	1116 kg/ha/yr	Sakhare,2015
Katepurna	Small	Maharashtra	223 kg/ha/yr	Sakhare,2015
Halai	Large	Madhya Pradesh	49.81 kg/ha/yr	Sakhare,2015
Umaim	Small	Meghalaya	65.52 kg/ha/yr	Sakhare,2015

Poaching

Poaching is a general problem in each and every reservoir. Fishery staff can prevent this, but there would be a large expenditure for preventing the poaching. Poachers can be punished under Indian Penal Code for theft or for offences committed in violation of rules made under the fisheries act of the respective state government. Legal measures for preventing poaching may not be effective in many states. In some reservoirs, it appears that poaching yields greater production than legal fishing.

Multiplicity of Ownership

Most of the reservoirs are administratively controlled by irrigation, revenue, forestry or electricity departments. These departments do not pay required attention towards development of fisheries. Naturally, Department of fisheries is not in a position to interfere in reservoir management. Therefore, the fishing ownership of all reservoirs should be transferred to respective state fisheries departments in all the states (Desai, 2008).

REFERENCES

Anon, 1979.Important fish production technologies developed by Central Inland Fisheries Research Institute for farmers of India, Central Inland Fisheries Research Institute, Barrackpore, West Bengal.

Anon,1999.Handbook of statistics, Department of Animal Husbandry and Dairying, Ministry of Agriculture, Government of India, New Delhi.

Desai, V.R.2008.Reservoirs can boost inland fish production. *Fishing Chimes*. 28(1): 39-46.

Dixitulu, J.V.H.2010.Why is the giant sleeping still? *Fishing Chimes*, 29(10): 5-6.

Gleick, P.H.1995.Water in crisis: A guide to the world's freshwater resources, Oxford University Press, Oxford, 1993, p.367; H. Chunhong. Cont rolling reservoir sedimentation in Chija, *The International Journal of Hydropower and Dams*, March 1995.

Jhingran, A.G.1988.Reservoir fisheries of India. *Journal of the Indian Fisheries Association*,18: 261-273.

Jhingran,A.G.1990.Recent Advances in Inland Fisheries Development inIndia.In: Technologies for Inland Fisheries Development (Eds.V.V.Sugunan and Utpal Bhaumik),Central Inlsnd Capture Fisheries Research Institute, Barrackpore, West Bengal,1-13.

Jhingran,V.G.1983.Fish and Fisheries of India,Hindustan Publishing Corporation (India),New Delhi.

Lalrinsanga, P.L., Asha Landge and Umashankar Prasad.2006.Stocking as a tool for inland fisheries management. *Fishing Chimes*, 26(2): 25-29.

Natarajan, A.V.1974. Planning strategies for development of reservoir fisheries in India (Abstract), CICFRI, Barrackpore, West Bengal.

NCA. 1976.Fisheries, Vol.8, National Commission on Agriculture, Government of India, New Delhi.

Rawson,D.S.1951.Total mineral content of lake waters.*Ecology*,32: 669-772.

Rawson, D.S.1955.Morphometry as a dominant factor in the productivity of large lakes.*Verh.Int.Ver.Limnol.*, 12: 164-175.

Ryder, R.A.1965.A method of estimating the potential fish production of North-temperate lakes.*Trans.Amer.Fish.Soc.*94: 214-218.

Sakhare, V.B.2007.Applied Fisheries, Daya Publishing House, Delhi.

Sakhare, V.B. 2015.Fish and Fisheries of Indian Reservoirs, Astral International Pvt. New Delhi.

Saxena, B.S.1990.Potentialities, production and problems of Indian reservoir fisheries,p.39-44.In: Jhingran,Arun,G. and V.K. Unnithan (Eds.).Reservoir Fisheries in India. Proceedings of the National Workshop on Reservoir Fisheries,3-4 January,1990.Special Publication 3,Asian Fisheries Society, Indian Branch, Mangalore, India.

Sinha, M.1998.Policy options for integrated development of reservoir fisheries: from production to marketing. *Fishing Chimes*. 18(1): 54-59.

Sreenivasan, A.1989.Principles of ecology and fisheries management of man-made lakes. In: Conservation and Management of Inland Capture Fisheries Resources of India (Eds. A.G. Jhingran and V.V. Sugunan), pp. 96-105, Inland Fisheries Society of India, Barrackpore, West Bengal.

Sreenivasan *et al.*, 1980.Development of reservoir fisheries in Tamil Nadu. *India Today and Tomorrow*, 8(4): 189p.

Srivastava, *et al.*,1985.Inland Fish Marketing in India,Vol.4a: 2,3,213 and 405.

Sugunan,V.V.1995.Reservoir Fisheries of India, FAO Fish.Tech.Paper No.345,FAO,Rome.

Vass, K.K.2006.Sustainable development of fisheries in inland waters of India. *Fishing Chimes*, 26(1): 188-191.

Yadava, P.K.1982.Reservoir Fishery Management: Major policy issues for government intervention. National Seminar on Fisheries Development in India, 9-11th April, Indian Institute of Management, Ahemadabad.

Chapter 16

Effect of Thyroxine on Growth of Fry of Rosy Barb, *Puntius conchonius*

☆ *R.K. Sadawarte, V.R. Sadawarte, J.M. Koli, G.N. Kulkarni, S.S. Sawate and K.M. Shinde*

ABSTRACT

Observations after 15 days of treatment showed no significant increase in body length and body weight in case of 2.5 mg/kg treatment. Where as there was a significant increase in the average body length ($p < 0.01$), body weight ($p < 0.001$) in the treatment of 5 mg/kg as compared to that of control. After 60 days of treatment there was a significant increase in the body length and body weight of both the treatments compared to that of control. The treatment using 2.5 mg/kg diet showed significant increase in the body length ($p < 0.01$) and body weight ($p < 0.001$) only after 60 days. In the treatment using 5 mg/kg diet, a significant increase was observed in body length ($p < 0.05$) and body weight ($p < 0.001$) after 60 days.

Keywords: *3-nitro, Growth, Rosy barb, Puntius conchonius.*

INTRODUCTION

It is well known that amphibian metamorphosis is induced by the thyroid hormone. Probably thyroxine plays a role in the parr – smolt transformation of salmonid fishes (Hoar, 1976; Bern, 1978; Dickhoff *et al.*, 1978). It was observed that the lower dosages of 0.025 to 0.075 mg were effective in reducing the hatching period, accelerating yolk absorption and improving the growth of postembryonic stages in *Brachydanio rerio, Cyprinus carpio, Labeo rohita, Cirrhinus mrigala* and *Catla*

catla (Sawant and Belsare, 1994).Thyroxine also plays a role in the acceleration of absorption of the yolk sac in *Sarotherodon niloticus* larvae (Nacario, 1983). It also accelerates the transition to free swimming in larvae of *Acquidens portalegrensis* (Munro,1984) and morphogenesis of the larvae of sturgeon, *Acipenser stellatus* (Iakavleva,1949) and chum salmon, *Oncorhynchus keta* (Dales and Hoar,1954). Thyroxine also accelerates growth and development in post yolk sac larvae of milk fish, *Chanos chanos* (Lam *et al.*, 1985). Thyroxine also enhances larval growth development and survival in *Sarotherodon mossambicus* and the carp, *Cyprinus carpio* (Lam and Sharma, 1985). Thyroid hormone is known to have its effect on the growth rates of fish (Donaldson *et al.*, 1978).

The application of thyroxine in the improvement of the larval rearing and accelerating growth rate in fishes, gave a start to the present investigation whether the use of thyroxine would be effective for the growth and early maturation in aquarium fishes. The present investigation was taken up to study the effect of thyroxine on the fry growth of Rosy barb (*Puntius conchonius*).

Materials and Methods

Brood stock of Rosy barb were obtained from the local aquarium fish dealers of Ratnagiri. They were reared in the laboratory in 90 x 20 x 30 cm size glass aquaria for required experimental stage. The average length of fry were 14.00 mm and average weight of fry were 26.00 mg for control and both group. The new born fry were initially fed with freshwater zooplankton and acclimated to test diet 4 days prior to the commencement of experiment. As a source of live feed, freshwater zooplankton was mass cultured using phased fertilization method (Shirgur, 1971) in plastic pool. Ten days old Rosy barb were reared in the laboratory, in 90 x 20 x 30 cm size all glass aquaria.

Preparation of Diet

A special flake diet was prepared in the laboratory using different local ingredients such as bone less fish meal (20 per cent), prawn shell waste (40 per cent) and wheat flour (40 per cent) (Table 16.1). The powdered dry ingredients were passed through 0.5 mm mesh size, then they were weighed and mixed together in the mixer. The dry mixture was mixed in water and blended for 5 min. in a blender. This moistened mixture was cooked at 80° to 90° C till the slurry gets the binding property. The slurry was then cooled to ambient temperature (29 ± 1°C).

Thyroxine @ 2.5 mg/kg and 5 mg/kg were dissolved separately in a small volume of distilled water. The insoluble substances were filtered out and the filtrate was collected and incorporated into the cooled slurry. The cooled slurry was spread on a polyethylene sheet with a smooth brush and sun dried and flakes were prepared. The flakes were packed in airtight containers and stored at room temperature and used whenever required. The control diet was without hormone.

A fortnightly record of the total length and weight of each fish from each group was kept from 1st day to 60 days. The average weight and length for each group of fish were recorded. For the experiment 20 fry was used for control group and each treatment group and fed with flakes twice a day.

Table 16.1: Different Doses of Thyroxine Incorporated Diets and their Ingredients

Sl.No.	Feed Ingredients	Control	Thyroxine (2.5 mg/kg)	Thyroxine (5 mg/kg)
		All values in (per cent)		
1.	Fish meal	20	20	20
2.	Prawn shell waste	40	40	40
3.	Wheat flour	40	40	40

(a) Fish meal (bone less) : Crude protein 70 per cent

(b) Prawn shell waste : Crude protein 7 per cent

(c) Wheat flour : Crude protein 12 per cent

Results

The observations on body length and body weight of each group are given (Table 16.2). The graphical representation of length and weight are also given (Figure 16.1).

Observations after 15 days of treatment showed no significant increase in body length and body weight in case of 2.5 mg treatment where as there was a significant increase in the average body length ($p < 0.01$), body weight ($p < 0.001$) in the treatment using 5 mg/kg as compared to that of control.

After 60 days of treatment there was a significant increase in the body length and body weight of both the treatments compared to that of control. The treatment using 2.5 mg/kg diet showed significant increase in the body length ($p < 0.01$) and body weight ($p < 0.001$) only after 60 days.

In the treatment using 5 mg/kg diet, a significant increase was observed in body length ($p < 0.05$) and body weight ($p < 0.001$) after 60 days.

Discussion

Thyroxine used in the dose of 0.1 ppm was found to enhance the larval development, growth and survival in *Sarotherodon mossambicus* 0.05 and 0.1 ppm in *Cyprinus carpio* (Lam and Sharma, 1985). Thyroxine plays an important role in acceleration of yolk sac absorption in *S. niloticus* larvae (Nacario, 1983), Chanos chanos (Lam *et al.*, 1985) is also reported.

It is not practically possible to treat young larvae by immersion in thyroxine solution. Therefore, in the present study thyroxine was incorporated into the diet. Higgs *et al.* (1982) suggested studies using thyroxine incorporation into diets.

The present study using thyroxin incorporated into the artificial diet was carried out in two experiments one using 2.5 mg thyroxin per kg diet and the other using 5 mg thyroxine per kg diet. The results obtained from the experiment using 2.5 mg thyroxine did not show significant increase in the length and weight upto 45 days of treatment. However after 60 days of treatment there was a significant increase in the length and weight of the young ones.

Table 16.2: Effect of Thyroxine on fry of *P. conchonius*

Sl.No.	Observations	Control		Thyroxine (2.5 mg/kg)		Thyroxine (5 mg/kg)	
		L (mm)	w (mg)	L (mm)	w (mg)	L (mm)	w (mg)
1.	0 Days	14.00±0.00	26.00±0.00	14.00±0.00	26.00±0.00	14.00±0.00	26.00±0.00
2.	15 Days	16.25±0.34	56.65±4.95	16.50±0.33	59.96±4.68	17.80**±0.33	80.85***±4.13
3.	30 Days	16.50±0.38	57.05±2.87	16.70±0.40	64.95±5.81	18.00**±0.36	88.48***±21.36
4.	45 Days	17.05±0.48	62.90±7.08	19.37±0.59	77.84±10.29	19.10**±0.38	103.53***±5.99
5.	60 Days	18.55±0.33	71.67±6.86	20.70**±0.63	133.3***±12.18	21.05*±1.24	154.41***±4.61

± Standard error of mean, * $P < 0.05$, ** $P < 0.01$,*** $P < 0.001$.

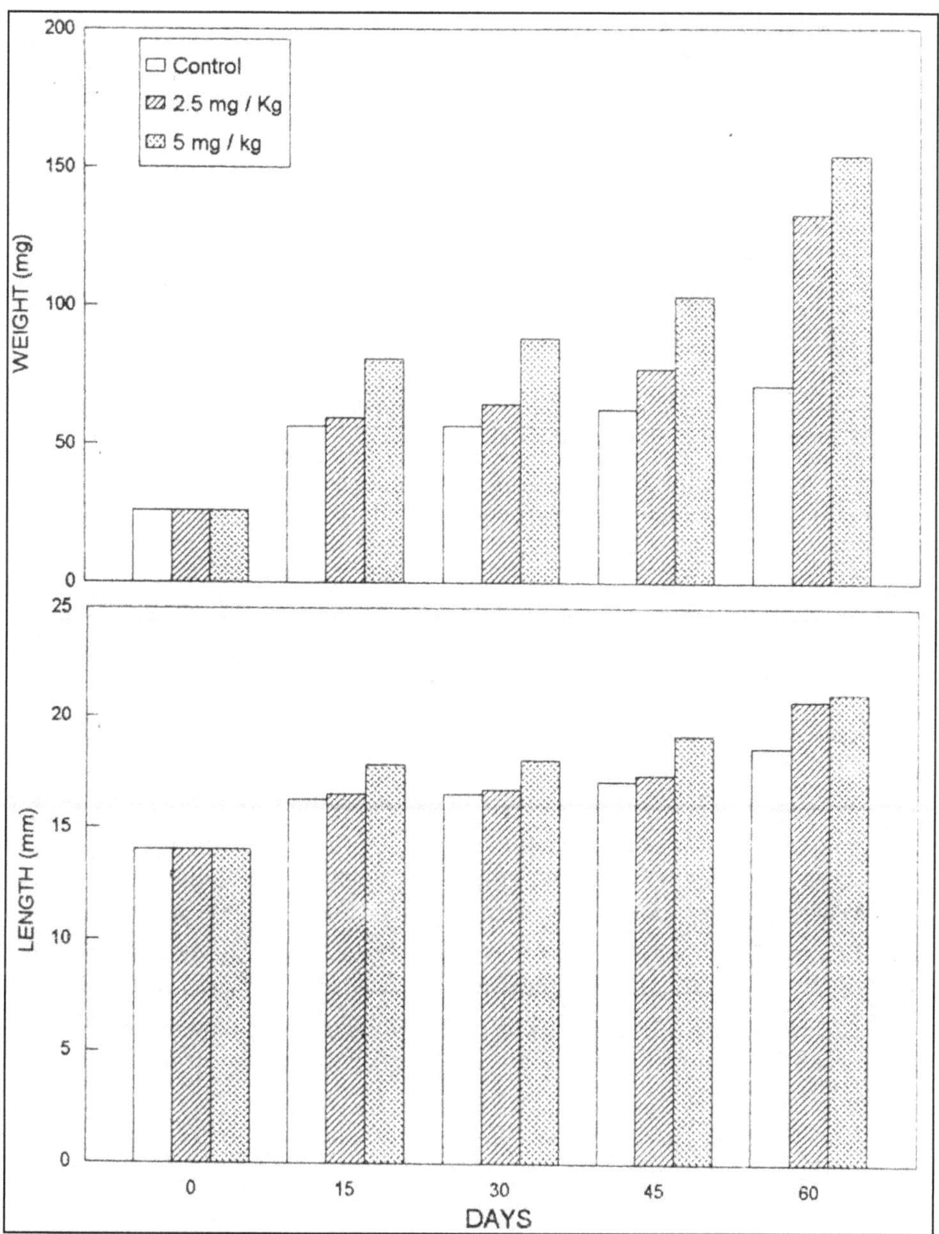

Figure 16.1: Effect of Thyroxine on Fry of *P. conchonius*.

In the experiment using 5 mg thyroxine there was a significant increase in the length and weight of the young ones after 15 days. Hence the treatment using 5 mg thyroxine diet was proved to be the best among the two.

Acknowledgements

Authors are highly obliged to Dr. B. Venkateswarlu, Hon'ble Vice-Chancellor, Dr. B.S. Konkan Agricultural University, Dapoli for facilities provided. Dr. P.C. Raje, Ex Associate Dean and Dr. R. Pai, Associate Dean, College of Fisheries, Ratnagiri for their keen interest and helpful suggestions.

REFERENCES

Bern, H.A., 1978. Endocrinological studies on normal and abnormal salmon smoltification. In : P.J. Gaillard and H. H. Boer (Editors), Comparative endocrinology. Elsevier/North-Holland-Biomedical press, Amsterdam, pp.97-100.

Dales, S. and Hoar, W.S., 1954.Effects of thyroxine and thiourea on the early development of chum salmon, *Oncorhynchus keta*. *Can. J. Zool.*, 32: 244-251.

Dickhoff, W.W., Folmar, L.C. and Gorbman, A., 1978.Changes in plasma thyroxine during smoltification of coho salmon, *Oncorhynchus kisutch*. *Gen. Com. Endocrinol.*, 36: 229-232.

Donaldson, E.M., Fagerlund, U.H.M., Higgs, D.A. and McBride, J.R.,1978.Hormonal Enhancement of Growth, In: W.S. Hoar; D.J. Randall and J.R. Brett (Editors), Fish Physiology, Vol., VIII. Academic Press, New York, NY, pp. 456-597.

Higgs, D.A., Fagerlund, U.H.M., Eales, J.G. and McBride, J.R., 1982. Application of thyroid and steroid hormones as anabolic agents in fish culture. *Comp. Biochem. Phisiol.*, 73B: 143-176.

Hoar, W.S., 1976.Smolt transformation : Evolution, behavior and physiology. pp.296.

Iakavleva, I.V., 1949.The independence of the activity of the thyroid gland from the thyrotropic function of the hypophysis in the post – embryonic development of Acipenserines. Dokl. Akad. Nauk., U.S.S.R., 60 : 281-284 (in Russian).

Lam, T.J., and Sharma, R., 1985.Effects of salinity and thyroxine on larval survival, growth and development in the carp, *Cyprinus carpio*. *Aquaculture*, 44 : 201-212.

Lam, T.J., and Juario, J.Y., and Banno, J., 1985.Effects of thyroxine on growth and development of post yolk sac larvae in milk fish, *Chanos chanos*. *Aquaculture*, 46: 179-184.

Munro, A.D.,1984. The ontogeny of the retina and optic tectum in *Aequidens portalegrensis* (Hensel). *J.Fish.Biol.*, 24 : 377-393.

Nacario, J.F.,1983.The effects of thyroxine on the larvae and fry of *Sarotherodon niloticus* (L). *Aquaculture*, 34 : 73 – 83.

Sawant, N.H. and Belsare S.G.1994.Effect of thyroxine on the hatching and postembryonic growth of *Brachydanio rerio*, *Cyprinus carpio*, *Labeo rohita*, *Cirrhinus mrigala*, and *Catla catla*. The silver jubilee of Indian Fisheries Association

National Symposium on Aquacrops organized by Indian Fisheries Association in collaboration with Central Institute of Fisheries Education (Indian Council of Agricultural Research) Versova, Mumbai, pp. 42.

Shirgur G.A., 1971.Observations on rapid production of zooplankton in fish nurseries by intensive phased manuring. *Ind. J. Fish.*, 1 (2) : 25-34.

Chapter 17

Gonadosomatic Index and Hepatosomatic Index of Freshwater Male Fish *Mystus cavasius* (Ham.) During Various Reproductive Phases in Bhadra Reservoir, Shivamogga, Karnataka

☆ *H.M. Ashashree, M. Venkateshwarlu and H.A. Sayeswara*

ABSTRACT

Influences of four reproductive phases in the testis (GSI) and liver (HSI) weight have been studied in a population of freshwater fish Mystus cavasius inhabiting Bhadra reservoir. Testes weight (GSI) was found to be maximum during prespawning phase where as liver weight was found to be minimum in the same phase. GSI is found to be directly proportional to prespawning phase and inversely proportional to post-spawning phase. The GSI and HSI provide an idea about the different reproductive phases of the present fish from the study area. In the males the change in GSI may influence the change in HSI too, as the role of liver in spermatogenesis appears.

Keywords: *Gonado Somatic Index (GSI), Hepato Somatic Index (HSI), Mystus cavasius, Bhadra Reservoir, Prespawning, Testes.*

INTRODUCTION

Maturity and spawning is an important biological aspect to be studied in fishes as it is useful in its management (Anju and Anoop, 2003). Fish is the major food resource for human being. The nutritive value of various species and types of fishes compares more favorable than muscles of beef, pork and mutton (Parulekar and Bal, 1969). Since fish represents a much larger source of potential proteins for human consumption than has been utilized, it seems more desirable to bring the most important facts relating to the food value of economically important species, on seasonal basis and in relation to important metabolic process of feeding, growth and breeding (Bumb and Parulekar, 2002).

The biological indexes of reproduction show the way in which fish use environmental and energetic resources. The Gonadosomatic index is a good indicator of reproductive activity, than being used in determining fish reproductive cycle stages (Devalming *et al.*, 1982). Variations in the hepatosomatic index of teleost are related to the liver capacity to store glycogen, physiological conditions, reproductive activity, feeding habits and food availability in their habitat (Dias *et al.*, 2000).

Mystus cavasius is commercially important, having a good price per kg in the market. The culture practice of this fish is not reported from this area. The gonado somatic index and hepato somatic index is one of the important parameter of the fish biology of Bhadra reservoir, which gives the detail idea about the fish production. Hence, the present study was under taken. Since gonado somatic index is an important tool in establishing the breeding period in animal and fish. The liver has a role in the testis development therefore; the HSI was correlated with GSI.

Several researchers have studied the biology of freshwater fishes (Ahirro, 2002; Sindhe and Kulkarni, 2004; Kulkarni, 2004; Shankar and Kulkarni, 2005; Shendge and Mane, 2006; Sunita *et al.*, 2011; Shailja and Saksena, 2012; Nzeh and Lawal, 2012).

Materials and Methods

The study was conducted during the period of December 2004 to November 2005. During the present investigation fishes were collected from Bhadra reservoir (13°- 42′-00″N latitude and 75°-38°-20″E longitude).The fishes were collected and brought to the laboratory and they were measured accurately for their total length, weight up to the nearest unit. Further, fishes were sacrificed after decapitation, length and weight of the fish and weight of the testes and weight of the liver were recorded. The morphological changes of the testes and liver were observed during different stages. The GSI and HSI were calculated by using the formula.

1. GSI

The gonado somatic index was determined by using the formula of Giese (1959).

$$GSI = \frac{\text{Weight of the testes}}{\text{Weight of the fish}} \times 100$$

2. HSI

The hepato somatic index was determined by following the formula of Gingerich (1982).

$$HSI = \frac{Weight\ of\ the\ liver}{Weight\ of\ the\ fish} \times 100$$

Statistical Analysis

A one-way analysis of variance (ANOVA) followed by Duncan's test ($P<0.05$) was performed to compare the means of the biological indexes for monthly reproductive cycle.

Results

The percentage of the gonad in the total weight of the fish is known as gonado somatic index and these values are an index of maturation of fishes (Varghese, 1971). In the present study, it has been observed that the values of gonado somatic index (GSI) and hepato somatic index (HSI) in the fish *Mystus cavasius* follow regular changes in the different months of four reproductive phases such as preparatory, pre-spawning, spawning and post-spawning phases. The GSI values vary from 0.2187±0.0186 to1.5836±0.0896. Where as the HSI values vary from 1.84±0.202 to 0.90±0.14.

Gonadosomatic Index (GSI)

The Gonadosomatic index of the *Mystus cavasius* was recorded during 4 different reproductive phases is presented in the Table 17.1 and depicted in Figure 17.1. The GSI increased during preparatory phase, reached maximum during pre-spawning phase, thereafter it reduced and reached minimum during post-spawning.

Table 17.1: Month-wise Values of GSI and HSI of male Fish *Mystus cavasius*

	GSI	*HSI*
January	0.38±0.05	3.62±0.01
February	0.59±0.04	2.45±0.09
March	0.77±0.01	2.09±0.09
April	1.15±0.02	1.26±0.05
May	1.39±0.01	1.24±0.03
June	1.45±0.09	1.10±0.01
July	1.64±0.02	0.80±0.11
August	1.44±0.05	1.05±0.01
September	1.36±0.01	1.69±0.01
October	1.29±0.11	3.00±0.02
November	0.61±0.03	3.99±0.03
December	0.34±0.01	5.95±0.08

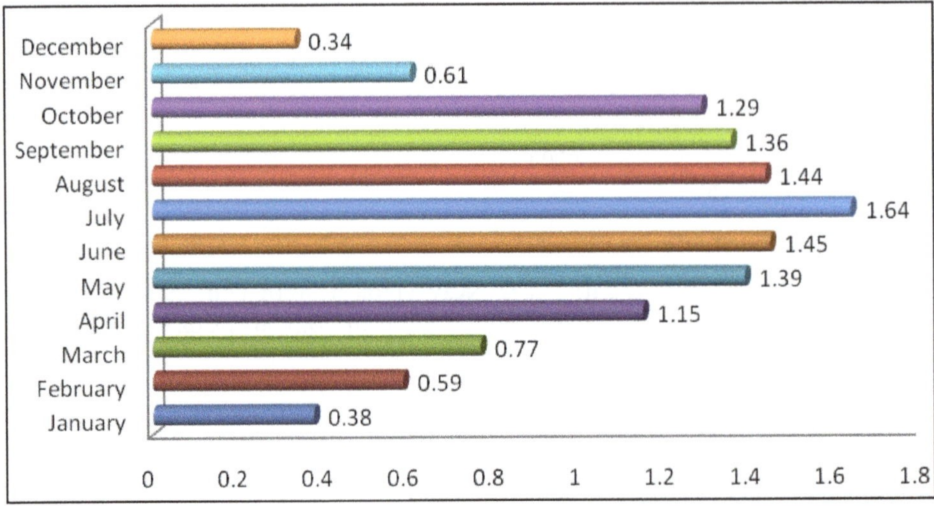

Figure 17.1: Monthly Variation in GSI of *Mystus cavasius*.

Hepatosomatic Index (HSI)

The HSI of fish studied during the 4 phases of reproductive cycle is presented in Table 17.1 and depicted in depicted in Figure 17.2. It indicates that maximum HSI was observed during post-spawning phase and the lowest HSI recorded during pre-spawning phase.

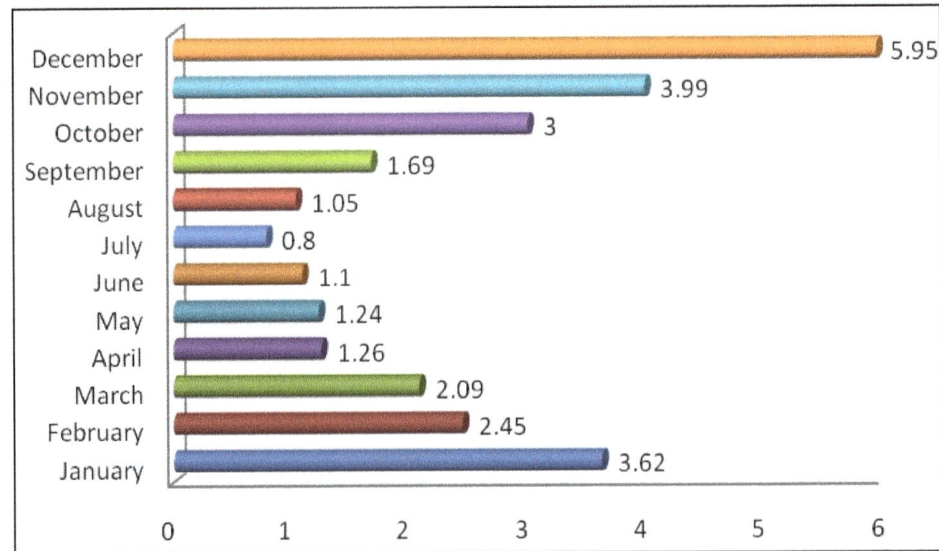

Figure 17.2: Monthly Variation in HSI of *Mystus cavasius*.

A comparison was made between GSI and HSI, shows that increase of GSI and decrease of HSI during pre-spawning phase indicates utilization of hepatic content for growth of testis suggesting their relationship. Comparison of monthly variation of GSI and HIS is shown in Figure 17.3. Changes of GSI and HIS during reproductive phases are give in Table 17.2 and shown in Figure 17.4. Similar response of reduction in HSI and comparable increase of GSI during spawning and reduction of GSI and increase of HSI during post spawning phase also indicated liver involvement in testicular growth and spermatogenesis.

Table 17.2: Changes in GSI and HSI in the Male Fish *Mystus cavasius* during Four Reproductive Phases

Reproductive Phases	*GSI*	*HSI*
Preparatory phase	0.58±0.08	2.72±0.005
Pre-spawning phase	1.39±0.03	1.1±0.09
Spawning phase	1.36±0.07	1.9±0.04
Post spawning phase	0.47±0.01	4.97±0.11

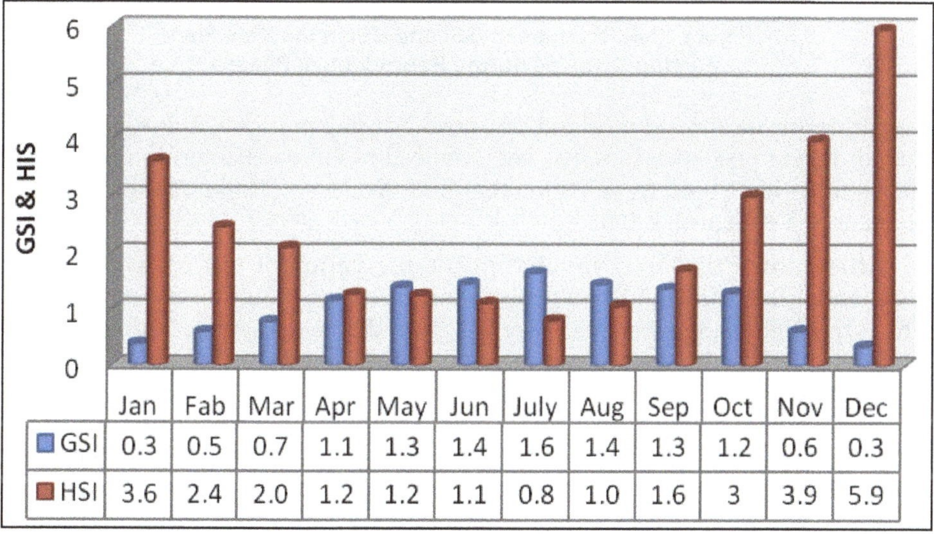

Figure 17.3: Comparision of Monthly Variation of GSI and HSI of *Mystus cavasius*.

Discussion

The percentage of gonad in the total weight of fish is known as gonado somatic index and this value is an index of maturation of fishes (Varghese, 1971 and Ahirrao, 2002). Singh and Singh (1979) also stated that the high values of GSI of fishes are the indicator of peak activity of gonads. The changes in the GSI have been considered as a sensitive parameter to monitor gonadal growth.

The gonadal cycle of *Mystus cavasius* can be divided into four phases during one year period as preparatory phase during January to March, pre spawning phase during April to July, spawning phase during August to October and post spawning phase during November to December. The developing gonad is in the

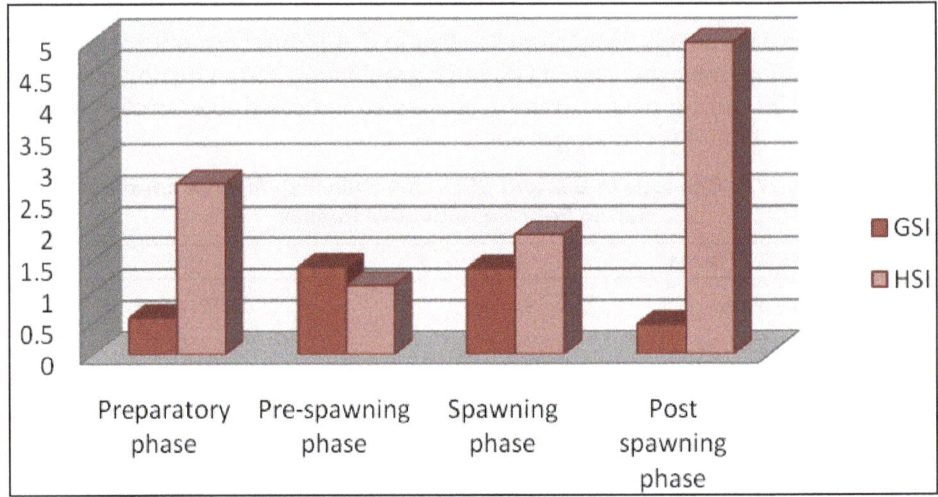

Figure 17.4: Changes in GSI and HSI in the Male Fish *Mystus cavasius* during Reproductive Phases.

preparatory phase, developed and ripe gonad during pre spawning and regressed gonad during post-spawning has been noticed based on changes in the GSI and such studies have been reported in other Indian teleosts (Malhotra *et al.*, 1989 and Shankar and Kulkarni, 2005).

It is known that liver involve in the regulation of testicular growth by contributing the biochemical components and providing energy for spermatogenesis. The correlative changes between liver weight and gonadal activity has been shown to be associated with energy requirement of the gonads for the development of spermatogenesis. It has been also shown that GSI and HSI have relationship and this relationship is directly related to gonadal activity. The results on comparison between GSI and HSI during different phases in *Mystus cavasius* in the present study shows that when GSI is increased the HSI is decreased during breeding phase of a fish and when GSI is decreased, HSI is increased during non-breeding phase.

According to Shankar and Kulkarni (2005) increasing GSI values are associated with maturation, whereas decreasing values are related to gamete extrusion. GSI was increased from immature to mature stage but declined at spent stage, whereas The HSI was increased during spent phase and preparatory phase, this high value of HSI indicates heavier liver. It indicates that, GSI and HSI have interrelationship and it may be because of the involvement of the liver in the seasonal growth of the testes. HSI is inversely proportional to GSI. The evaluation of the HSI must take the role of both endogenous and exogenous factors. The HSI varies with seasonal cycles because of the role in storage and metabolic activities, nutritional quality also affect relative liver size (Shankar and Kulkarni, 2005). Cambray and Bruton (1984) have suggested the HSI of the fish *Babus anoplas* lowering of Somatic condition factor following spawning was due to spawning activity and cessation of feeding during spawning. Singh and Singh (1979) found that the high values of HSI during preparatory and post-spawning and low levels during pre-spawning and spawning in the *Heteropneustis fossils*.

The GSI values corresponding to spermatogenetic activity in the testes of different teleosts have been observed. Bhatti and Al-Daham(1978), Thakur (1978), Llewellyn (1979) and Shankar and Kulkarni (2005) suggests that, the reduction in the condition of HSI during spawning and post spawning indicates exhaustive condition after breeding activity in fish *Notopterus notopterus* and cessation of feeding during spawning. So a majority of teleost fishes all over the world are seasonal breeders and in the Indian sub continent a vast majority of the freshwater fishes breed during the heavy rain fall of monsoon months. (Jhingran, 1982). Environmental factors are well known tools, influence reproductive maturity (Cambray, 1994; Chadhuri, 1997).

In the fish *Mystus cavasius* spawning occurs during monsoon periods. Hence this fish can be categorized as monsoon spawners (Malhotra *et al.*, 1989). Similar observations were also made in *Padogobius sarana sarana* (Patil and Kulkarni, 1996); in *Macrozoarces americanus* (Wang and Crim, 1997); in *Padogobius martensi* (Cinquetti and Dramis, 2003).

Conclusion

In the reproductive phases of *Mystus cavasius*, the GSI increased with the progressive development of the gonads until the gonads become ripe and the index declined sharply in the spawning and the spent fishes. Therefore it can be concluded that in the present fish *Mystus cavasius*, both GSI and HSI have inter-relationship and it may be due to the involvement of liver in the seasonal growth of the testes.

REFERENCES

Ahirro, S.D. 2002. *Status* of gonads in relation to total length and GSI in freshwater spiny eel *Mastacembelus armaatus* (Lacepede) from Marathwada Region, Maharashtra. *J. Aqua. Biol*, 17 (2): 55-57.

Anju Thapliyal, Anoop, K. D. 2003. Maturation and spawning biology of a hill stream catfish *Pseudecheneis* (Mc clelland) from the Garhwal Himalaya, Uttaranchal, *Ad. Bios.* 22 (1): 9-22.

Bhatti, M.N and Al-Daham, N.K. 1978. Annual cyclical changes in the testicular activity of a freshwater teleost, *Barbus luteus* (*Heckel*) from shaft Al-Arab, *Iraq.J. Biol.*, 13: 321-326.

Cambray, J.A., Bruton, M.N. (1984). The reproductive strategy of a barb, *Barbus annnopllus* colonizing in a manmade lake in SouthAfrica, *J. Zool.* 204: 143-168.

Cinquetti, R., Dramis, L. 2003. Histological, histochemical, enzyme histochemical and ultra structural investigation of the testis of *Padogobius martensi* between annual breeding seasons, *Journal of Fish Biology*, 63: 1402-1428.

De Vlaming, V.L., Santos, G.B., Chapman, F. 1982. On the use of the gonosomatic index. *Comparative biochemistry and physiology, Vancouver* 73 A (1): 31-39.

Jhingran, V.G.1982. Fish and fisheries of India. Hindustan Publishing Corporation (India), Delhi. (1982).

Llewellyn, L.C.1979. Some observation on the spawning and development of the Mitchellian freshwater hardyhead, *Craterocephalus fluviatilis,* Mc Culloch. *Austral. Zool.*, 20: 269-288.

Malhotra, Y.R, Youth, M.K., Gupta 1989. Reproductive cycles of freshwater fishes. In: reproductive cycles of Indian vertebrates (Saidapur.S.K. *Ed*).Allied publishers Ltd., New Delhi.pp. 58-105 (1989).

Nzeh, C.G., Lawal, A.2012. Condition Factor, Gonadosomatic Index and Sex ratio of the Family Mormyridae from a small Lake in Horin, Nigeria. *World Journal of Zoology*.7 (2): 102-105.

Parulekar, A. H., Bal, D.V.1969. Observations on the seasonal changes in the chemical composition of *Bregraceros mc Cleelandi* (Tnomoson), J. *Univ. Bombay* XXXVII. 65: 88-92.

Shailja Mishra, Saksena, D.N. 2012. Gonadosomatic index and fecundity of an Indian major carp *Labeo calbasu* in Gohad reservoir. *The Bioscan*.7 (1): 43-46

Shankar, D.S., Kulkarni, R.S.2005. Somatic condition of the fish,*Notopterus notopterus* (pallas) durilng different phases of the reproductive cycles, *J. of Environmental Biology*. 26 (1): (2005).

Shendge, A.N., Mane, U.H. 2006. Gonadosomatic Index and spawning Season of Cyprinid fish *Cirrhina reba* (Hamilton), *J.Aqua. Biol*, 21 (1): 127-129.

Sindhe, V.R., Kulkarni, R.S. 2004. Gonadosomatic and hepatosomatic indices of the freshwater fish *Notopterus notopterus* (Pallas) in response to some heavy metal exposure, *Journal of Environmental Biology*, 25(3): 365-368.

Sunita,K., Kulkarni, K.M.,Gijare, S.S., Tantarpale, V.T.2011. Seasonal changes of Gonadosomatic indiex observed in the Freshwater fish *Channa punctatus. The Bioscan*. 6(4): 571-573.

Thakur, N.K. 1978. On the maturity and spawning of an air - breathing catfish *Clarias batrachus* (Linn.), *Matsya*, 4: 59-66.

Varghese, T.J.1971. Studies on maturation, spawning and sex ratio of *Coilia ramcarati* Gunther, *J. Indian Fish. Asso*. 1(1): 58-67.

Wang, Z., Crim, L.W.1997. Seasonal changes in the biochemistry of seminal plasma and sperm motility in the ocean pout, *Macrozoarces americanus*, *J. of Fish Physiology and Biochemistry*, 16: 77-83.

Chapter 18

Length Weight Relationship of Shovel-Nosed Lobster *Thenus unimaculatus* (Burton and Davie, 2007) from Veraval Fishing Habour

☆ *V.M. Chavda, J.B. Solanki, P.V. Parmar, H.V. Parmar and V.C. Bajaniya*

ABSTRACT

Scallarid lobster Thenusunimaculatusis monotypic species contributing to the commercial catches in Gujarat. Incidental catches in the mechanized trawler along the coast brings concerns about the Thenus unimaculatus. The study on length weight relationship of Thenus unimaculatuswas carried out from Veraval fishing harbour for a period of Dec. 2013 to Nov. 2014. Length-weight equations estimated for Thenus unimaculatuswas LogW = -4.3934 + 2.8912 LogLand LogW= -4.5247 + 2.9638 logL for males and females respectively.The present study revealed the proportionate increase in length and weight in Thenus unimaculatus.

Keywords: *Length-weight relationship, Shovel-nosed Lobster, Thenus unimaculatus.*

INTRODUCTION

Thenus unimaculatus is a decapod crustacean belong to the family Scyllaridaewith feeding habit of carnivorous scavenger preferring benthic invertebrates includes molluscs, polychaetes and crustaceans. The characteristics of the genus Thenus is very well described and several works (Holthuis 1991, Burton and Davie 2007). The Thenus has been reported to be bottom dwelling species occurring at sandy or muddy substratum. (Uraiwan 1977, Jones 2007, Kizhakudan *et al.*, 2004, FAO 2010).*Thenus unimaculatus* is available along both the east and west coasts of India and contributes to the commercial fisheries, where their landings are mainly as by-catch of trawlers. Study of length-weight relationship has a significant role in fishery biology and has several applications, since various important biological aspects *viz.*, general well-being, onset of maturity and spawning, fecundity *etc.* can be assessed with the help of condition factor from this relationship (Le Cren, 1951). The present study is aimed to find out the biological parameters of data deficient species *Thenus unimaculatus* off the Veraval coast.

Material and Methods

A total of 483 specimens of *T. unimaculatus* were observed for length and weight from Veraval fishing harbour during December 2013 to November 2014 from trawl landings. Total number of 205 males and 278 females were recorded for total length and total wet weight in the present study.The parameters of the length weight relationship were obtained by fitting the power function, $W=axL^b$(LeCren,1951) where W is the total weight, a is constant determined empirically, L is the total length.

Results and Discussion

The regression coefficients of the length-weight relationship of *T. unimaculatus* were analyzed for the period of one year from December 2013 to November 2014. The length-weight relationship of males and females during the study period were analyzed with least square method (Table 18.1). The monthly data were pooled for one year and regression equation was obtained following.

$$\text{LogW} = -4.3934 + 2.8912\ \text{LogL}$$

The correlation co-efficient (r) value for male is 0.98 ($p<0.001$).

$$\text{LogW} = -4.5247 + 2.9638\ \text{logL}$$

The correlation co-efficient (r) value for female is 0.97($P<0.001$).

Table 18.1: Observations Recorded for *T. unimaculatus*

Total number of specimens analyzed	385
Length range observed	62mm to 230 mm
Weight range observed	8 gm to 323 gm
Length weight relationship for male	LogW = -4.3934 + 2.8912 LogL
Length weight relationship for female	LogW= -4.5247 + 2.9638 logL

In the present study, the regression line (Figures 18.1 and 18.2) shows a close relationship between males and females. The values of b were 2.8912 for males and 2.9638 for females which is negative allometric ($b<3.0$). In both the cases value of r found to be greater than 9.0. The present study revealed the proportionate increase in length and weight in *T. unimaculatus*. Similar results were found by Kabli and Kagwade (1996), Subramanian (2004), Soumendra Nath *et al.* (2009).

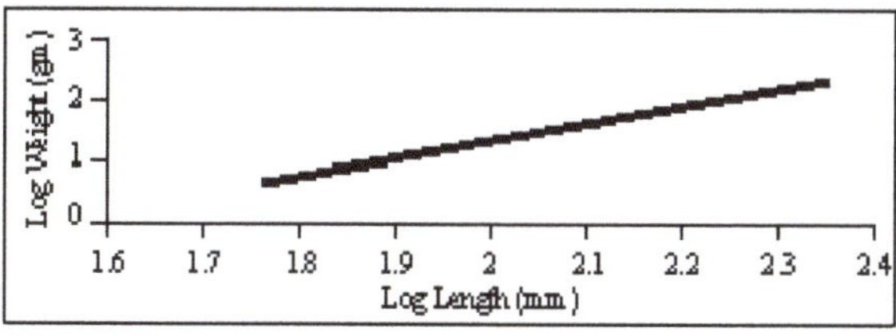

LogW = -4.3934 + 2.8912 LogL

Figure 18.1: Length-Weight Relationship in Males.

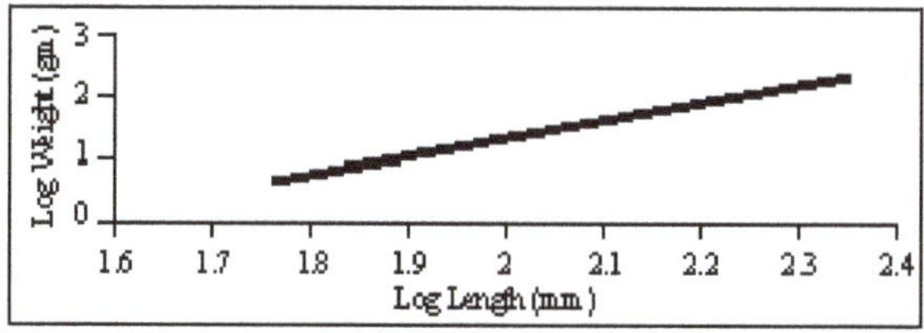

LogW = -4.3934 + 2.8912 LogL

Figure 18.2: Length-Weight Relationship in Females.

To make an effective and viable fishery it is essential that the biology of the animal is comprehensively understood. In the face of overexploitation of fishery resources, it is necessary to make conservation efforts to demersal benthic species such a *T. unimaculatus* which shows slow growth rates and exploitation of brooders. In Mumbai, the slipper lobster T. orientalis disappeared from the fishery by 1994 (Deshmukh, 2001) due to their slow growing and relatively low fecundity (Kabli and Kagwade, 1996).

Figure 18.3: *Thenus unimaculatus.*

REFFRENCES

Burton TE, PJF Davie. 2007. A revision of the shovel-nosed lobsters of the genus *Thenus* (Crustacea: Decapoda: Scyllaridae), with descriptions of three new species. *Zootaxa* 1429: 1-38.

Deshmukh, V. D. 2001.Collapse of sand lobster fishery in Bombay waters. *Indian J. Fish.*, 48: 71-76.

FAO. 2010. Fishery statistical collections: global production.Food and Agriculture Organization (FAO) of the UN.

Holthuis LB. 1991. FAO species catalogue vol. 13: marine lobsters of the world. An annotated and illustrated catalogue of species of interest to fisheries known to date. FAO Fisheries Synopsis 125 (Supplement 13): 292.

Jones CM. 2007. Biology and fishery of the bay lobster, *Thenus* spp. *In* KL Lavalli, E Spanier, eds. The biologyand fisheries of the slipper lobster. *Crustacean Issues,* Vol. 17. Boca Raton, FL: CRC Press, pp. 325-358.

Kabli, L.M. and P. Kagwade, 1996.Morphometry and conversion of the sand lobster, *Thennus orientalis* (Lund) from Bombay waters. *Indian J. Fish.,* 43: 249-254.

Kagwade.P.V. and L.M. Kabli 1996. Ageand growth of the sand lobster *Thenus orientalis* (Lund) from Bombay water. *Indian J. Fish.,* 43(3): 249-251.

Kizhakudan, Joe K., P. Thirumilu, S. Rajapackiam and C.Manibal. 2004. Captive breeding and seed productionof scyllarid lobsters - opening new vistas in crustaceanaquaculture. *Mar. Fish. Infor. Serv. T and E Ser.,* 181: 1-4.

Le Cren, E.D., 1951. The length-weight relationship andseasonal cycle in the gonad weight and conditionin the perch (*Perca fluviatilis*). *J. Anim. Ecol.,*20: 201-219.

Subramanian v. (2004).Fishery of sand lobster *Thenus orientalis* (Lund) along Chennai coast. *Indian J. Fish.,*51(1): 111-115.

Soumendra Nath S *et al.,* 2009. Length-weight relationship and relative condition factor in *Thenus orientalis* (Lund,1893) along East Coast of India. *Current Research Journal of Biological Sciences.* 1(2): 11-14.

Uraiwan, S. 1977. Annual Report. Biological study of *Thenus orientals* in the Gulf of Thailand. Bangkok, Thailand: Department of Fisheries.

Chapter 19

Catch Composition of Non-Motorised Gillnet Fishery Operating in Mumbai Coast

☆ *Shabir A. Dar, Saly N. Thomas, S.K. Chakraborty, M.H. Balkhi and Farooz A. Bhat*

ABSTRACT

A total of 19 different fish and shell fish varieties were landed by non-motorised gillnetters at Mahim of Mumbai coast. The Catfishes contributed 16.69 per cent of the total landing. Among three groups, pelagic fishes shared 49.61 per cent, demersal fishes 43.81 per cent while crustacean contributed only 7.59 per cent of the total landing. Among the total pelagic catch oil sardine contributed 32.40 per cent while as in demersal fin fishes, catfishes contributed maximum (38.10 per cent) and in crustacean group only crabs were observed with a share of 7.59 per cent to the total catch. The fish varieties and their percentage contribution varied from month to month. Maximum landing was observed in September contributing 15.52 per cent of the annual catch.

Keywords: *Gillnet, Catch composition, Mumbai coast, Maharashtra.*

INTRODUCTION

Gillnetting is a common fishing method practiced by mechanised and artisanal fishermen in the marine, estuarine and freshwater environment. Gillnets are passive fishing gears, being vertical walls of netting kept erect in water column by means of floats and sinkers and set perpendicular to the direction of movement of target fish (Hameed and Boopendranath, 2000)."Gillnetting" the name itself is self-explanatory which means fishes are caught by gilling. Gillnetting is a popular passive fishing technique with ancient origin. Gillnets primarily catch the fish by entanglement of the gill cover in a mesh bar. The mesh size is just large enough to pass head of the fish but not the rest of the body. When the fish tries to swim through the mesh, it feels the tightening of the mesh around its body and swims back when its operculum gets entangled into the mesh bar and the fish gets gilled. Snagging, wedging and entangling are the other ways by which fishes are caught in this type of net. The capture of fish in gillnets depends on the net construction, its dimension and shape of the particular species of fish encountered (Thomas, 2009). Sardine, seer, pomfret, mackerel, ilisha, shark, catfish, shrimp and crab are the common varieties of fish and shell fish caught in gillnets.

The earliest evidence of gillnet came from herring drift nets operated in the North Sea in 11th and 12th centuries. It is the most important traditional gear practised and is a low energy fishing technique common in inland and inshore waters of the world. In world fisheries, gillnets (20 percent) rank next to trawl and purse seines in terms of total catch (Kellehar, 2005). Over the years, there have been dramatic developments in the fishing systems. Passive gears generally being energy efficient have gained importance because of economic pressure to reduce fuel costs and they are expected to gain increasing acceptance in the years ahead.

Though gillnets of Maharashtra play a prominent role in the marine fish landings of the state, a comprehensive study on this fishery with special focus on catch composition, has not been done. Since 1980's many need based changes have taken place mainly with respect to material substitution, introduction of resource specific gear, use of coloured webbing, and other changes consequent to this. Hence, the present study was undertaken.

Materials and Methods

Fishing season in Maharashtra starts on the day of *Narali Pournima* or on 15th of August, whichever is earlier whereas it is closed before onset of monsoon or generally on 10th June. Fishing activities remain suspended in most of the landing centres in this region during the monsoon months as the fishermen find it difficult to navigate their vessels due to rough weather. In addition to this, as per Maharashtra Marine Fishing Regulation Act, 1981, there is a ban on fishing in monsoon season from 10 June to 15 August or up to *Narali Pournima,* whichever is earlier. So data were collected from 1st December 2010 to 30th November 2011 except during the fishing ban period.

Catch statistics from non-motorised gillnet sectors from Mahim fish landing centre of ten boats were recorded by weekly observation in terms of number of baskets or crates or weight in kilogram landed. The mesh sizes of gear used on the particular day were determined by measuring stretched mesh with a centimeter scale (FAO, 1978). The data on species wise/group wise and total catch were collected as per Alagaraja (1984). Identification of the catch upto species level was made following Fisher and Bianchi (1984).

Results and Discussion

Catch Composition

During the period under study it was observed that non-motorised gillnetters were mostly operating nylon monofilament gillnets of mesh sizes 26-32 mm for anchovies, sciaenids and crabs, 48-50 mm for catfishes and 70-75 mm for solefish at a depth of 2.5 to 10 m.

Species/Group-wise Catch Composition

A total of 19 different fish and shell fish varieties were landed by non-motorised gillnetters at Mahim of Mumbai coast from December 2010 to November 2011.

Catfishes contributed 16.69 per cent of the total landing, followed by oil sardine (15.99 per cent), croakers (15.99 per cent), crabs (7.59 per cent), other shads (7.27 per cent), Bombay duck (1.43 per cent), Indian mackerel (6.13 per cent), anchovies (5.30 per cent), horse mackerel (3.55 per cent), hilsa (4.06 per cent), sole (4.03 per cent), mullets (3.52 per cent), terapon (3.50 per cent), silver bellies (1.54 per cent), thread fins (1.09 per cent) whereas, remaining groups *viz.* scads, wolf herring, pomfrets and seerfish together was less than one per cent and total contribution is 2.55 per cent of the total landing catch.

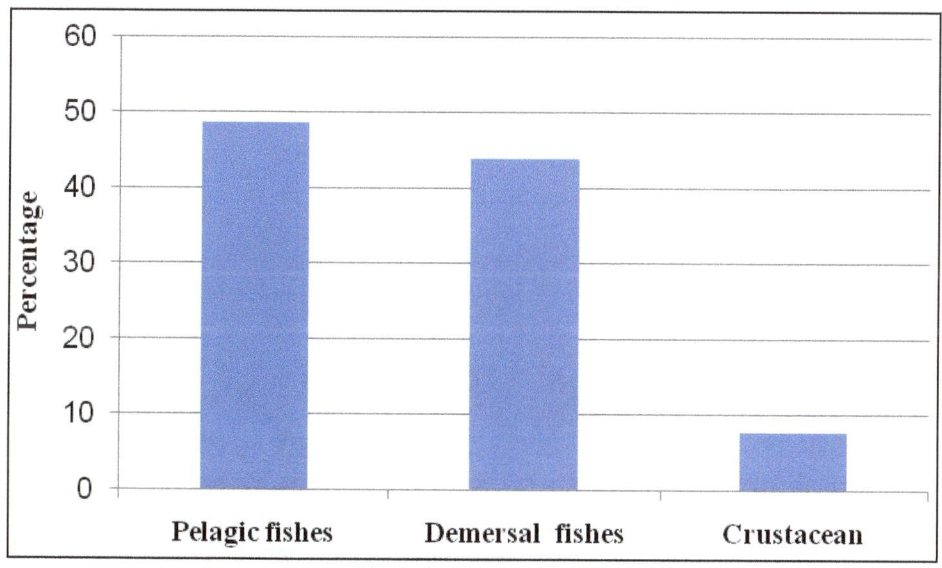

Figure 19.1: Percentage-wise Contribution of Various Resources to the Total Catch of Non-motorised Gillnetter.

The average landing of non-motorised gillnetter for one fishing season operated from Mahim fish landing centre was 1,792 kg in which pelagic fishes were 871 kg, demersal fishes were 785 kg and crustacean 136 kg shown in Figure 19.1. Among three groups, pelagic fishes shared 49.61 per cent, demersal fishes 43.81 per cent while crustacean contributed only 7.59 per cent of the total landing. Oil sardine contributed 32.40 per cent of the total pelagic catch followed by other shads (14.96

per cent), Indian mackerel (12.62 per cent), anchovies (10.90 per cent), horse mackerel (7.31 per cent), hilsa (8.35 per cent), mullets (7.23 per cent), Bombay duck (2.94 per cent), scads (1.87 per cent), and seerfish 0.61 per cent of the total catch. Among demersal fin fishes catfishes contributed maximum (38.10 per cent) followed by croakers (36.49 per cent), solefish (9.19 per cent), terapon (8.00 per cent), silverbellies (3.51 per cent), and threadfins (2.48 per cent) and in crustacean group only crabs were observed with a share of 7.59 per cent to the total catch.

Month-wise Catch Composition

Maximum landing was observed in September contributing 15.52 per cent of the annual catch. The October month contributes 10.24 per cent, November (9.56 per cent), December (10.24 per cent), January (9.66 per cent), February (8.03 per cent) and March (7.58 per cent). Minimum landing was recorded in April with 7.48 per cent contribution to the total catch and May (7.80 per cent) (Figure 19.2). Hilsa, other shads, anchovies, catfishes, sole and crabs were observed throughout the study period. Croakers and terapon were recorded throughout the year except in March and November respectively. The results indicated that there was no variation in landings between months ($P>0.05$) whereas there was significant difference in species composition among months ($P<0.01$).

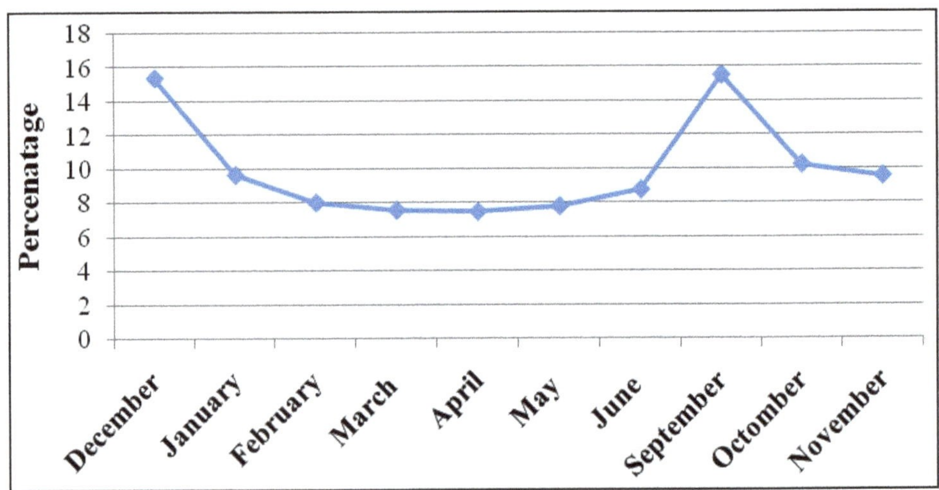

Figure 19.2: Month-wise Percentage Contribution to the Total Landing of Non-motorised Gillnetter.

Catch composition of any fishing operation gives an idea about the varieties of fish and shellfish availability in that region, which in turn helps in better understanding of the biodiversity. In addition to this type, of fish quality and quantity landed by any gear play an important role in its economic viability.

In non motorised gillnetters catfishes was the major landing followed by oil sardine and croakers. Group wise catch composition was also taken into consideration while analysing the catch data to get an idea about particular groups, dominance in the total catches. In non-motorised gillnetters total landing for one fishing season was 1,792 kg in which pelagic landing was 871 kg by demersal fish landing was 785 kg and crustacean 136 kg.

Seerfishes are exploited mainly by the gillnets, as 65.1 per cent of the all India seerfish catch during 1989-94 and 1995-99 periods is by gillnets (Devaraj *et al.*, 1999, Muthiah *et al.*, 2003). Drift gillnets contributed 13.3 per cent of the goatfishes landed during 1998-2000 and 10 per cent of the total sciaenid catch during 2000 (Mohanraj *et al.*, 2003). In Vizhinjam, Kerala, 63 per cent of the sardine catch was contributed by gillnets (Lazarus and Thiagarajan, 1994).

Along the south east and north east coasts 45 per cent and 64.9 per cent respectively of whitefish (*Lactarius lactarius*) was contributed by this gear during 1991-99 (Vivekanandan *et al.*, 2003a). At Tuticorin, gillnet is the major gear catching lesser sardines of the 1- year and zero year class (Pillai and Rohit, 2003). In Tamil Nadu, 31 per cent of the total catch are contributed by gillnets (Manisseri and Radhakrishnan, 2003). At Mumbai, 50.2 per cent of the pomfrets landed during 1997-98 to 2000 was contributed by gillnets (Sivakami *et al.*, 2003).

Thus, it can be concluded that varieties of fish and shellfishes caught all along the Indian coast by gillnet varied from place to place. Species reported by Sehara and Karbhari (1989) along the coast of Maharashtra and by Koya and Vivekanand (1992) along the Veraval coast were in accordance with the catch composition of the present study. Since the area of their study was very close to that of the present study.

Gillnet is a selective gear and its mesh size is designed to catch a particular size class of fish species. Gillnets with various mesh sizes were operated along different parts of the country (Silas *et al.*, 1984; Sehara and Karbhari, 1989; Sathiadhas *et al.*, 1991; Koya and Vivekanandan, 1992; Thomas *et al.*, 2005) to catch different species and sizes of fish hence the variation in catch composition.

Anon, (2013) estimated that gillnets contributed golden anchovy (11.07 per cent), Indian mackerel (5.8 per cent), tuna (10 per cent), barracuda (13.5 per cent), wolf herring (46 per cent), sharks (77 per cent), pomfret (46 per cent), and catfish (4.01 per cent) to the total marine catch landed during 2012 along Maharashtra coast.

In non motorised sector most productive month was September constituting 15.37 per cent of the total landing followed by December (15.37 per cent) and October (10.24 per cent). Silas *et al.* (1984) reported July contributing 19.43 per cent of the gillnet catch as the most productive month followed by August (18.7 per cent) and May (12.07 per cent). In the present study, data during July was not collected due to ban period and thus landings of this month were not recorded.

Similar findings were also reported by Sehara and Karbhari (1989) at selected centres along the Maharashtra coast. They had reported that the post-monsoon quarter (September-November) was the most productive with seerfish as major contributor. Sathiadhas and Panikkar (1988) reported monsoon as the most productive season along the Trivandrum coast, whereas, Koya and Vivekanandan (1992) have reported maximum landings in September from Versova coast.

During the present study non-motorised groups October, November, December and January were the most productive months *viz.*, winter season was the most productive. Thus it can be concluded that the monsoon and the post-monsoon season are more productive than pre monsoon. This may be due to effect of upwelling occurring on the west coast of India during monsoon period, which brings in nutrients and increases the productivity of the coastal area, which has played a significant role in fish production with regard to gillnetter (Rajagopalan *et al.*, 1992). Thus, the catch composition of gillnetters operated along Mumbai coast varied with reference to the mesh size of the nets used for operation.

REFERENCES

Alagaraja, K. 1984. Simple methods for estimation of parameters for assessing exploited fish stocks, *Indian J. Fish.*, 31: 177-207.

Anon., 2013. *Annual Report,* Published by Central Marine Fisheries Research. Institute Cochin, during 2012 – 2013. pp. 1 - 200.

Deveraj, M., Kasim, H. M., Muthiah, C. and Pillai, N. G. K. 1999. Assessment of the Exploited Seer Fish Stocks in the Indian waters, *J. Mar. biol. Ass.* India, 4: 62-84.

FAO 1978. Catalogue of Fishing Gear Designs. Farnham, Surrey, Fishing News Books for FAO, 159 pp.

Fischer, W. and Bianchi, G. 1984. *FAO Species identification sheets for Fishery Purposes Western Indian Ocean* (Fishing Area 51), FAO, Rome.

Hameed, M. S and Boopendranath, M. R. 2000. Modern Fishing Gear Technology, Daya Publishing House, Delhi: 186 pp.

Kellehar, K. 2005. Discards in the World's marine fisheries. An update, *FAO Fish. Tech. Pap.*, Rome. 470: 131 pp.

Koya, K. P. S. and Vivekanandan, E. 1992. Gillnet fishery off Veraval during 1982-1990. *Mar. Fish. Infor. Ser.*, 116: 1- 4.

Lazarus, S. and Thiagarajan, R. 1994. An Assessment of the Exploitation of the Sardine Stock West Coast of India. *J. Mar. Biol. Asso*c. India, 36: 100-105.

Manisseri, M. K. and Radhakrishna, E. V. 2003. Marine Crabs, In: Joseph, M. M., Jayaprakash, A. A. (Eds.). Status of Exploited Marine Fishery Resource of India, Central Marine Fisheries Research Institute, Cochin, 188-194.

Mohanraj, G., Batcha, H. and Gomathy 2003. Sciaenids, In: Joseph, M. M., Jayaprakash, A. A. (Eds.). Status of Exploited Marine Fishery resource of India, Central Marine Fisheries Research Institute, Cochin, 133-140.

Muthiah, C., Pillai, V. G. K. and Kasim, B. 2003. Seer Fish, In: Joseph, M. M Jayaprakash, A. A. (Eds). *Status of Exploited Marine Fishery Resource of India,* Central Marine Fisheries Institute, Cochin, 45-48.

Pillai, N. G. K., and Rohit, P., 2003. Lesser Sardines In: Joseph, M. M., Jayaprakash, A. A (Eds.) Status of exploited Marine Fishery Resource of India, Central Marine Fisheries Institute, Cochin, pp. 25-29.

Rajagopalan, M. S., Thomas, P.A., Mathew K.J., Deniel Selvaraj G. S, George, R. M., Mathew C.V, Naomi T. S., Kaldharan P., Balachandran V. K. and Geetha, A. 1992. Productivity of Arabian Sea along the South West coast of India. *Bull. Cent. Mar. Fish. Res. Inst.* 45: 9-37.

Sathiadhas, R. and Panikkar, K. P. P. 1988. Socio-economics of small scale fishermen with emphasis on cost and earnings of traditional fishing units along Trivandrum coast, Kerala- A case study. *Seaf. Export J.*, 20: 21-37.

Sathiadhas, R., Benjamin, R. E. and Gurusamy, R. 1991. Technological options in the traditional marine fisheries sector and impact of motorization on the economics of gillnet fishing along Tuticorin coast, Tamil Nadu. *Sea f. Export J.*, 23: 26-36.

Sehara, D. B. S. and Karbhari, J. P. 1989. Gillnet fishing by mechanised boats at selected centres in Maharashtra and its profitability. *Seaf. Export J.* 21: 10-23.

Silas, E. G., Pillai, P. P., Jayaprakash, A. A. and Pillai, M. A. 1984. Focus on small scale fisheries: Drift gillnet fishery off Cochin. 1981 and 1982. *Mar. Fish. Infor. Serv.*, 55: 1-12.

Sivakami, S., Vivekanandan, E., Raje, S. G., Shobha, J. K. and Rajkumar, U. 2003. Lizard fishes. Pomfrets and Bullseye, In: Joseph, M. M., Jayaprakash, A. A. (Eds.). Status of Exploited Marine Fishery Resource of India, Central Marine Fisheries Research Institute, Cochin, 141-157.

Thomas, S. N. 2009. Selectivity of gillnets, In: Meenakumari, B., Boopendranath, M. R., Pravin, P., Thomas S. N. and Edwin, L. (Eds.). Handbook of Fishing Technology, Central Institute of Fisheries Technology, Cochin, 146-151.

Thomas, S. N., Meenakumari B., Pravin, P. and George Mathal, P. 2005. Gillnets in Marine Fisheries of India. Monograph, *Agriculture Technology Centre,* Central Institute of Fisheries Technology, Cochin, 45 pp.

Vivekanandan, E., Zacharia, P. U. and Mahadevaswamy, H. S. 2003. White fish, In: Joseph, M. M., Jayaprakash, A. A. (Eds.). Status of exploited Marine Fishery Resource of India, Central Marine Fisheries Research Institute. Cochin, 171-175.

Chapter 20

Socio-economic Status and Scope for Improvement of Navibandar Fishing Village of Saurashtra

☆ *V.M. Chavda, J.B. Solanki, H.V. Parmar, P.V. Parmar, V.T. Brahmane and V.M. Solanki*

ABSTRACT

The socio-economic survey was carried out at Navibandar fishing village in the Saurashtra region of Gujarat stateto evaluate the socio-economic condition of fishermen families and to find out means and scope for development of the fishing village. This investigation was done during April to May 2013.The fishing village was surveyed with personal interview, direct and indirect observation method. In this investigation, the population was surveyed for socio-economic status, family size, income and occupational patterns, age and gender structure, education, customs and beliefs, religion, standard of living, village problems, fishing techniques, fish catch composition, financial aspects, and role of women. Recommendations have been proposed for the improvement in socio-economic condition of fishermen families and overall growth of the village through some change in policies of government and community, diversification of occupation and development of entrepreneurship with the judicial utilization of natural resources available in the village.

Keywords: *Socio-economic status, Fishing village, Navibandar.*

INTRODUCTION

The marine fishing villages in India are mostly socially and economically backward. However, there are some advantage with the small scale fisheries associated with fishing village like Navibandar as they require less capital and operational cost and optimize human power with the use of more passive gears like gillnets and hence lower ecological impact.These kind of rural areas are more suited for small-scale fisheries since they are more labour-intensive and therefore can provide higher employment opportunities. Similar studies have been carried out by workers (Sathiadhas and Venkataraman, 1983;Sehra *et al.*, 1986) for about socio-economical aspects of fishermen in different regions of India.

Navibandar fishing village is endowed with potential coastal resources in the Saurastra region. Fish and fishing is the only the source of food, employment and the sole contributor to the economy of the village. The main objective of the study is to examine the socio-economic condition of fishermen families and to find out means and scope for development of the fishing village.

Material and Methods

The data collection was carried out in Navibandar fishing village located at 21°26′ N and 69°48′ Ewith direct and indirect observation, personal interview with structured questionnaire containing both open ended and closed ended questions. Since it is a small village, all the 193 families were surveyed for the present study. The data collected were family size, income and occupational patterns, age and gender structure, education, customs and beliefs, religion, standard of living, village problems, fishing techniques, craft and gear,fish species caught and its value, financial aspects, and role of women. This investigation was done during April to May 2013.

Results and Discussion

Navibandar fishermen village is at 30 km distance from Port Porbandar. Ninety percentage of population of village belongs to Kharvacaste.

Family Size

Families are nuclear family and normally 2-3 children have found in every family. Male and female childare equally represented in the families.

Income and Occupational Patterns

Majority of the population belongs to fisherman families, majority are engaged in fishing but some got a government jobs like driver, teacher, police, and Din Dyal ration store. Monthly income of fishermen family is Rs. 5000-7000.

Table 20.1: Age and Gender Structure, Population and Sex Ratio

Male	Adults	278
	Children	337
	Subtotal	615
Female	Adults	248
	Children	331
	Subtotal	579
Grand total		1194
Sex ratio	100 Female and 106 male	

Education

The adult generation is having primary education up to 7th standard. In case of old age above 50 there are literate but not learned more than 4th standard. The village owns primary and secondary school. Majority of village children discontinue their studies after secondary school and engaged in fishing activities.

Customs and Beliefs, Religion

Navibandar village followHindu religion and majority of the population belongs to Kharva community. Fishermen start fishing season only after worshiping fair of the Bhadraigodess. The fishing is self-suspended by the community during Hundu festivals like Egyarasi, Ekadasi, Ramnavmi, Diwali, Rakshabandhan, Bhaibij, Holi *etc.*

Standard of Living

The majority of the houses are made of concrete. Each house is accomplished with Toilet facilities are by Aga Khan, the NGO. Electricity connections are present in all houses. Houses are equipped with Television connected with DTH. Kitchen are equipped with stove with Gas cylinders. Main food of the community is fish and Bajra roti.

Fishing Techniques

In this village OBM FRP boats are using for in fishing,8-9 HP Yamaha engine. One-day fishing is carried out up to 5 Nautical miles off the coast in 20-30 feet depth zone. Gill net is used for fishing. During off-season they also use other gears like cast net,carb-traps and hook and line for fishing in the estuary. Using alternate gears like traps has been reported in other parts of India during monsoon seasons during which fishing in marine waters is not possible (Sehra *et al.*, 1992). Gill net of different mesh size is used according to type of species and season. During post-monsson season for the capture of "Pomfret Jadajal", a type of gill net is used. During winter season "Tangi", a type of gill net is used for capture of Seer fish. During season start after monsoon using 18-20cm size of gill net forpomfret fishing. During winter uses crocker gillnet size 8-9cm. postwinter season For seer fishing 9-12 cm in size gill net use.Normally monofilament and nylon are basic types of twine in gill net found

Fish catch per boat per trip is about 20 to 30 kg of fish in which Croaker and Mackerel species forms dominant groups.

Financial Aspects

The commercial banks are away 14 km from the Navibandar village. People are invest money in postal savings. SakhiMandal, women's cooperative society self-help programmeis actively working in village. SakhiMandal is a group of women which are provided facility of debit and credit along them women members. Danga, the middlemen, provide credit to fishermen as an advanced payment to the fish catch. In return fishermen have to sell their catch irrespective of its value to the Danga at low price.

Table 20.2: Species caught and its market value

Fishes Catches	Species Caught Scientific Name	Value Price of Rupees/kg
Yellow crocker	*Johniusdussumieri*	70-85
Silver crocker	*Johnieposspp.*	50-60
Seer fish	*Scomberomorusguttatus*	60-90
Wolf herring	*Chirocentrusspp.*	60-70
Pomfret	*Pampusargentus*	350-550
Cat fish	*Arius sp*	50-60
Eel	*Congresoxtalabonoides*	40-60
Ribbon fish	*Trichuruslepturus*	60-80
Sting rays	*Dasyatiszugei*	50
Mackerel	*Alepes mate*	30-50
Dog shark	*Scoliodonlaticaudus*	60
Mullet	*Mugilcephalus*	40-50
Tarpon	*Theraponjarbua*	30-40
Shrimp	*Penaeusmonodon*	300-400
Crab	*Scylla serrata*	
	Portunussanguinolentus	75-90

Role of Women

A few womenparticipate to sell the fish in nearby villages and Porbandar. Majority of fisher women carry the high value fish to their home and store it in chilled form in their houses with the use of ice. Many women are also helping mending damaged fishing gear. The role of women in fisheries at Navibandar fishing village is limited to postharvest storage of the raw material till for a short time, till the family get better price for the fish. The studies carried out east and west coast of India about the role of women in fisheries indicate that they actively participate in all the postharvest activities (Immanuel *et al.*, 2003). The present study indicate the limited role of women in fisheries indicate fishermen family's overall higher living standard.

Village Problems

The Village do not have primary health center and fishermen have to seek medical advice from nearby Gosa village or Porbandar city. Estuarine system of the village is open only during high tide, and it is the only time when fishermen can set off for fishing in the sea, during low tide it is not possible to enter or leave the creek, situation can become worst during adverse climate. With increasing population of the village and due to dependence only on artisanal nature of the fishing activity the unemployment is increasing the village. The village can increase the fish production as well as improve its social and economic conditions through recommendations provided here and can accomplish the greening in fisheries.

Strategies to Greening Fisheries of Navibandar Fishing Village

The fish resources yet are not harmed at the present level of fishing effort, but before it could deteriorate the ecosystem, diversification in fisheries activities can increase fish production and generate employment in the village without harming the resources at long run due to over capitalization. Attention can be provided for training of fishermen for diversification of their fisheries occupations.

The village is bestowed with clear and pollution-free brackish-water in the creek. Fishermen frequently capture brooders of commercially important varieties of shrimps like *Penaeusmonodon*. The coastal waters also harbor commercially important Oyster species of *Saccostreacucullata*. Microcredit can be provided to the fishermen for the investment in establishment of mini-hatcheries, cage culture, penculture, sea-weed culture etc through which the village can become self-sustaining system at a very low cost.

The elevation of the areas near creek is potential site for aquaculture activities. This land could have been allotted to the fishermen for aquaculture practices. The village is located near the National Highway No. 8E.The road transport facility to the village is quiet good hence none of the additional investment will be required for the infrastructure development for transportation.

REFERENCES

ImmanuelS., V.N. Pillai, E. Vivekanandan, K.N. Kurup and M. Srinath.2003.A Preliminary Assessment of the Coastal Fishery Resources in India – Socio-economic and Bio-economic perspective. p. 439 - 478. In: G. Silvestre, L. Garces, I. Stobutzki, M. Ahmed, R.A. Valmonte- Santos, C. Luna, L. LachicaAliño, P. Munro, V. Christensen and D. Pauly (eds.) Assessment, Management and Future Directions for Coastal Fisheries in Asian Countries. World Fish Center Conference Proceedings 67, 1 120 p.

Sathiadhas, Rand G. Venkataraman 1983. Indebtedness and utilization of fisheries credit in Sakthikulangara- Neendakara, Kerala. - A Case study. *Mar. Fish. Info. Serv T and E Ser.*, 54: 1-6.

Sehra, D. B., J. P. Karbhari and R. Sathiadhas 1986. A study on the socio-economic conditions of fishermenin some selected villages of Maharashtra and Gujarat coasts. *Mar. Fish. Info. Serv.*, 69: 1-18.

Sehra, D.B.S., K. K. P. Panikkar and J. P. Karbhari 1992. Socio-economic aspects of the monsoon fisheries of the west coast of India. *Bull. Cent. Mar. Fish. Res. Inst.*, 1992, 45: 242 – 250.

Chapter 21

Conservation and Mangement of Catfishes of Perumal Lake, Cuddalore District, Tamil Nadu

☆ *M.V. Radhakrishnan*

ABSTRACT

The inland fishers and rural community depend for their livelihood and food security on the available indigenous species. The present investigation has been conducted to conserve and manage the catfishes of Perumallake from October 2014 to September 2015 at weekly intervals. In the present study 13 species of catfish belonging to four families were collected from Perumallake. A maximum of seven species were recorded in the Bagridae family followed by three species in Clariidae and two species in Siluridae. A maximum of six species of catfish were noted as most abundant (Clarias batrachus, Heteropneustus fossilis, Mystus armatus, Mystus seengtee, Mystus malabaricus, Mystus bleekeri). The abundant species include Clarias gariepinus, Mystus menoda, Mystus oculatus, Ompok bimaculatus and Wallago attu. Vulnerable species is represented by Mystus gulio and endangered species (ED) is Clarias dussumieri.The results of the present study clearly showed that the fish fauna of the River is highly diverse and proper management is essential for the conservation of the fish biodiversity.

Keywords: *Perumallake, Catfish, Conservation, Management, Tamil Nadu.*

INTRODUCTION

India is endowed with rich fishery resources with great biodiversity and the inland fishers and rural community depend for their livelihood and food security on these indigenous species. Fish account for the highest species diversity among all vertebrates and they live in almost all conceivable aquatic habitats (Remadevi, 2003). The Ichthyofauna of the world constitute about 21,730 species of which about

2500 species are found in India (Jayaram,2009), 930 live in freshwater and harbors 197 species of catfish (Jayaram 2009). In recent years much interest has developed in the study of the phylogeny and taxonomy of the order Siluriformes as a whole (Jayaram 2009). The catfish diversity has been studied in seven species in Karala River, West Bengal (Patra, 2011), ten species in the wetlands of Shivamogga District, Karnataka (Kiran, 2011) and 17 species in River Kelo and Mand in Raigharh District, Chhattisgarh, India (Tamboli and Jha, 2012). Kharat *et al.* (2012) reported six species of catfish, out of 51 species of freshwater fishes from Krishna River at Wai and Dhom reservoirs. Dahanukar *et al.* (2012) reported 11 species of catfish, out of 57 species of freshwater fishes of Indrayani River, a tributary of Bhima River. Jadhav *et al.* (2011) reported 10 species of catfish of a total 58 species. Hence, the present study has been designed to for the conservation and management of catfishes of Perumallake Cuddalore district, Tamil Nadu, India

Materials and Methods

Study Area

Perumal Lake (11 30′ to 11 0 45′ N and 79 30′ to 79 47′ 30″ E) is located in Cuddalore district of Tamil Nadu and is bounded by Gadilam river in the north and Coleroon river in the south (Figure 21.1). The lake is being used for multipurpose utility such as irrigation, fish catching, washing and bathing.

Specimen Collection

Fish specimens were collected from the lake at weekly intervals for a period of one year from October 2014 to September 2015. Sampling was carried out using drag net of having mesh size 2mm with mesh size 3m x 1m size and also with cast net of standard size and identification of fishes was done (Day, 1978; Talwar and Jhingran, 1991; Tekriwal and Rao, 1999; Jayaram, 2009; Froese and Pauly, 2012).

Results

In the present study 13 species of catfish belonging to four families were collected from Perumallake. The scientific name, family, collection sites and status of each catfish species is given in Table 21.1. A maximum of seven species were recorded in the Bagridae family followed by three species in Clariidae and two species in Siluridae. (Table 21.1). A maximum of six species of catfish were noted as most abundant (*Clariasbatrachus, Heteropneustus fossilis, Mystus armatus, Mystus seengtee, Mystus malabaricus, Mystus bleekeri*). The adundant species include *Clarias gariepinus, Mystus menoda, Mystus oculatus, Ompok bimaculatus* and *Wallago attu*. Vulnerable species is represented by *Mystus gulio* and endangered species (ED) is *Clarias dussumieri* (Table 21.1).

Discussion

Fisheries that exploit a range of species may have more stable catches than fisheries that exploit a single species. Conservation of biodiversity is important in environmental management programs (Reid and Miller, 1989;Turner and Gardner, 1991).The environmentally sustainable use of fish resources is central to fisheries

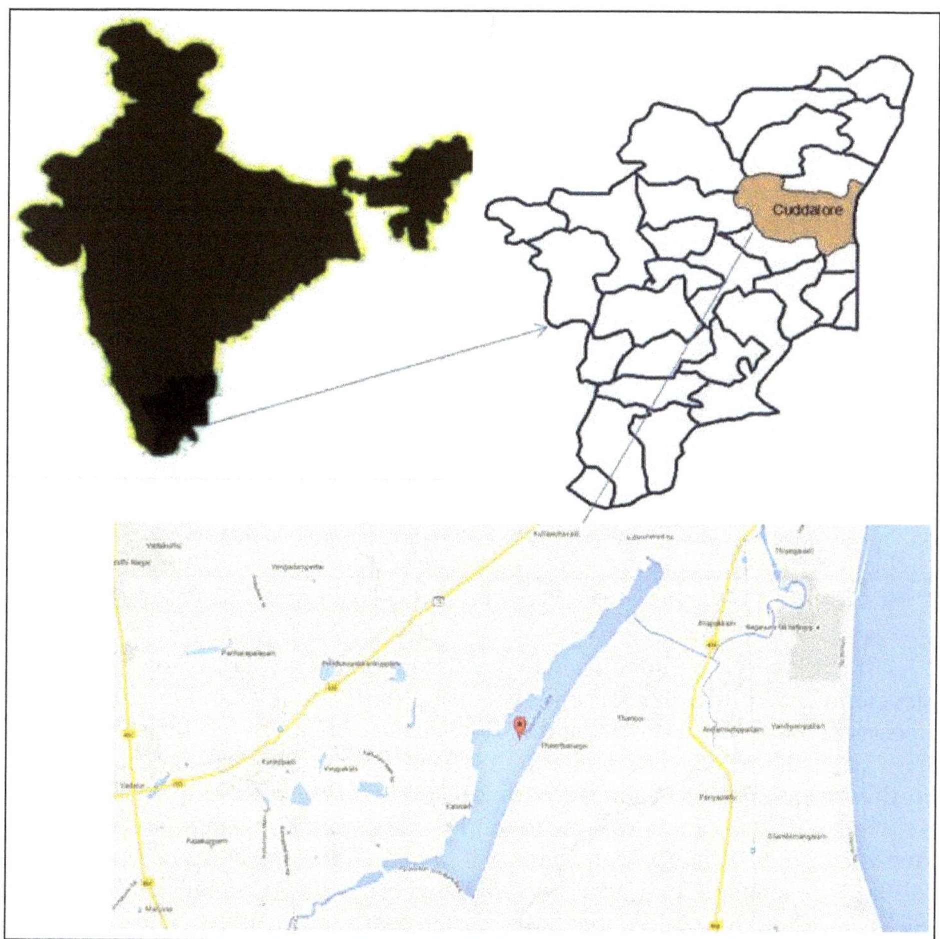

Figure 21.1: Perumal Lake.

management, given the long-term importance of this sector in terms of nutrition and employment. But today's major concern relates to the unsustainable levels of exploiting fishes with such practices that lead to the depletion of fish stocks, disruption of ecological equilibrium and reduction in diversity (Radhakrishnan and Sugumaran, 2012; Vijayasree and Radhakrishnan, 2014; Jithesh and Radhakrishnan, 2014). Measuring biodiversity is one of the central issues in ecology because of its importance in devising conservation strategies. Moreover, fisheries sector occupies a very important place in the socio-economic development of the country. It has been recognized as a powerful income and employment generator as it stimulates growth of a number of subsidiary industries, and is a source of cheap and nutritious food besides being a foreign exchange earner. Most importantly, it is the source of livelihood for a large section of economically backward population of the country. There is, therefore, need to throw light on sustainable use of small indigenous fish species and conservation of biodiversity.

Table 21.1: Catfishes of Perumal Lake

Sl.No.	*Genus and Species*	*Family*	*Status*
1.	*Clarias batrachus*	Claridae	**
2.	*Clarias dussumieri*	Claridae	ED
3.	*Clarias gariepinus*	Claridae	*
4.	*Heteropneustes fossilis*	Heteropneustidae	**
5.	*Mystusarmatus*	Bagridae	**
6.	*Mystus bleekeri*	Bagridae	**
7.	*Mystus gulio*	Bagridae	VL
8.	*Mystus malabaricus*	Bagridae	**
9.	*Mystus menoda*	Bagridae	*
10.	*Mystus seengtee*	Bagridae	**
11.	*Mystus oculatus*	Bagridae	*
12.	*Ompok bimaculatus*	Siluridae	*
13.	*Wallago attu*	Siluridae	*

**: Most Abundant; *: Abundant; –: Rare; VL: Vulnerable; ED: Endangered; CED: Critically Endangered.

The fish species is threatened by habitat modifications caused by dams, rapid development in urbanization and increasing pollution (Dahanukar 2011). Three species *M. malabaricus, W. Attu* and *O. Bimaculatus* are considered Near Threatened. The major threat is overexploitation of these species for food and has resulted in drastic population declines (Patra 2011). Threats also include anthropogenic activities such as, habitat loss due to sand mining, pollution and indirect mortality due to dynamite fishing in many parts of it's range (Abraham 2011; Ng 2010; Ng *et al.*, 2010). One exotic species *C. gariepinus* was also observed in Perumallake. This species might have been introduced into the lake from nearby aquaculture tanks/ponds and river The distribution of catfish in the lake may be due to the depth of water level that remains at 2–3 m even in summer which ensures the continuous availability of food and shelter while in monsoon the level rises up to 10 –12 m. The slow and steady water movement and its width ensure the continuous availability of nutrition. In addition, the bottom of the lake has rich fertile clay with varied aquatic flora and fauna. Hence, the present study suggests that Perumallake can be considered as a suitable habitat for the conservation of catfish. The lake is polluted due to discharge of sewages from the villages. Furthermore, unscientific practices such as dynamite fishing for collection of food fishes and sand mining are alarmingly increased. If the present trend continues, the adverse conditions might lead to the loss of the fish fauna. Since the catfish form commercially important food fish and this study suggests that the lake is a more suitable habitat for the conservation of freshwater catfishes, there is a need to take up conservation measures in order to increase the fish population in the perumal lake. Hence, to maintain a healthy biodiversity and abundance, the following activities/conservation measures be put in place; (i) do not harvest the fish during the spawning period, (ii) do not harvest juveniles, (iii) do not allow the introduction of invasive species such as tilapia, mosquito fish and guppies into the river and (iv) educate locals about the life cycles of freshwater fish and the negative impact of pollution with sewage, fertilizers, pesticides and other chemicals (Radhakrishnan and Sugumaran, 2012).

Acknowledgements

The author is thankful to the Dr. R. Karuppasamy, Professor and Head, Department of Zoology, Annamalai University, for the encouragement.

REFERENCES

Abraham, R.2011. *Mystusmalabaricus*. In: IUCN 2012. IUCN Red List of Threatened Species.Version 2012.2. http: / /www.iucnredlist.org Downloaded on 30 March 2013.

Dahanukar N. 2011. *Glyptothoraxpoonaensis*. In: IUCN 2012. IUCN Red List of Threatened Species.Version 2012.2.<http: / /www.iucnredlist.org>. Downloaded on 21 March 2013.

Dahanukar, N, M. Paingankar, R.N. Raut, and S.S. Kharat.2012. Fish fauna of Indrayani River, northern Western Ghats, India.*Journal of Threatened Taxa* 4(1): 2310–2317; http: / /dx.doi.org/10.11609/JoTT.o2771.2310–7.

Day,Francis. 1978.Collection of Indian fishes. Bulletin of the British museum (Natural History), *Historical Series*, 5(1): 1-189.

Froese,R., Pauly, D. 2012. Fish Base. World Wide Web electronic publication. www. fishbase.org, version, (08/2012).

Jadhav BV, SS Kharat, N Rupesh, R Raut, N Paingankar, N Dahanukar 2011. Freshwater Fish Fauna of Koyna River, Northen Western Ghats, India.*Journal of Threatened Taxa* 3(1): 1449–1455;http: / /dx.doi.org/10.11609/JoTT.o2613.1449-55.

Jayaram, K.C.2009. Catfishes of India.Narendra Publishing House. Delhi, India, 383.

Jhingran, V.G (1991). Fish and Fisheries of India, Edn 3rd. Hindusthan Publishing Corporation, India. 954.

Jithesh,M, MV Radhakrishnan.2014. On the fishes of Chaliyar River, Kerala State, India, *Ecology and Fisheries*, 7(2): 17 – 24.

Kharat, S.S., M. Paingankar, and N. Dahanukar.2012.Freshwater fish fauna of Krishna River at Wai, northern Western Ghats, India.*Journal of Threatened Taxa* 4(6): 2644–2652; http: / /dx.doi.org/10.11609/JoTT. o2796.2644-52.

Kiran, B.R.2011. Diversity, distribution and abundance of Murrels and Catfish resources in some wetlands of Karnataka, India.*Asian Journal of Experimental Biological Sciences* 2(2): 349–353.

Ng, H.H.2010. *Wallago attu*. In: IUCN 2012. IUCN Red List of Threatened Species. Version 2012.2.<http: / /www.iucnredlist.org>. Downloaded on 21 March 2013.

Ng, H.H., K Tenzin, and M. Pal.2010.*Ompokbimaculatus*. In: IUCN 2012. IUCN Red List of Threatened Species.Version 2012.2.<http: / /www.iucnredlist.org>. Downloaded on 30 March 2013.

Patra, A.K.2011. Catfish (Teleostei: Siluriformes) diversity in Karala River of Jalpaiguri District, West Bengal, India. *Journal of Threatened Taxa* 3(3): 1610–1614; http: / /dx.doi.org/10.11609/JoTT.o2474.1610-4.

Radhakrishnan,M.V.and Sugumaran, E.2012. Fish diversity as a cue to the fisheries potential of veeranam lake, Tamil Nadu, India. Paul VI (ed.) In Biodiversity: Issues Impacts, remediation and Significances. VL Media Solutions Publishers, New Delhi, 162-166.

Reid, W.V.and Miller, K.R.1989. Keeping Options Alive: The scientific basis, Springer, New York.

Remadevi, K.2003. Freshwater fish biodiversity. In Venkataraman K (Ed) Natural Aquatic ecosystems of India, Zoological Survey of India, Chennai : 217- 224.

Talwar, P.K. and Jhingran, V.G.1991. Inland Fishes of India and adjacent countries. In 2 vols. Oxford and IBH Publishing House, New Delhi, 1158.

Tamboli, R.K. and Jha,Y.N.2012. Status of catfish diversity of River Kelo and Mand in Raigarh District, CG, India.*ISCA Journal Biological Scences*1: 71–73.

Tekriwal, K.L. and Rao, A.A.1999. Ornamental Aquarium Fish of India. Waterlooville, Kingdom Books : 144.

Turner, M.G.and Gardner, R.H.1991. Quantitative measures in landscape ecology, conserving biodiversity of Washington: World Research Institute.

Vijayasree, T.S., Radhakrishnan M.V.2014.Fish Diversity of Kuttanad River, Kerala State, India. *International Journal of Fisheries and Aquatic Studies.* 1(6): 55-58.

Chapter 22

Construction of FRP Trawler: A Recent Technique

☆ *G.S. Temkar, S.Y. Metar, S.B. Satam and N.D. Chogale*

INTRODUCTION

Fibre-glass boats are extremely strong, and not rust, corrode or rot over the period of time. Fibre-glass provides structural strength, especially when long woven strands are laid, sometimes from bow to stern, and then soaked in epoxy or polyester resin to form the hull of the boat. Whether hand laid or built in a mould, FRP boats usually have an outer coating of gel-coat which is a thin solid colored layer of polyester resin that adds no structural strength, but does create a smooth surface which can be buffed to a high shine and also acts as a protective layer against sunlight. People have even made their own boats or water craft out of materials such as foam or plastic, but most of them in present situation are built of plywood and either painted or covered in a layer of fibre-glass and resin.

Around the mid 1960s, boats made of glass-reinforced plastic, more commonly known as fibre-glass, are used, specially for recreational boats (Shamsuddin, 2003). The United States Coast Guard refers to such boats as 'FRP' (for Fibre Reinforced Plastic) boats. Fibre-glass reinforced plastic (FRP) is a rather new material in boat building. It is suitable for building vessels of various types, ranging from 2 m rowing boats up to an 80 m naval minesweeper (Coackley *et al.*, 1991) with an estimated hull-life of about 20 years (Laszlo, 1960). The first introduction of FRP in the boat building was in the late 1940s and was used for the military purposes in North America (Coackley *et al.*, 1991). The first fishing vessel was built in 1967, operating in the North Sea and in the United States of America; the first FRP trawler was built in the early 1970s and is still in service.

Methodology and Data Collection

For collection of data, interview of the carpenters was taken from the FRP construction boat building yard. Personal interview of 10 members was taken from the yard. Information about construction, materials used, chemical used, expenditure *etc.* was gathered through interview.

Results and Discussion

1. Construction of Mould

First inner surface of mould is cleaned to make it free from dust (Figure 22.1). Then make a mixture of PVA (*Polyvinyl alcohol*) and wax. That mixture was useful for removal of mould easily after complete drying of boat structure. Then apply that mixture on the mould. Kept it for half an hour, allow to dry.

Figure 22.1: Mould Structure for Boat Construction.

2. Preparation of Gel-Coat Mixture with Colour

Then mixture of gel-coat (coloured layer of polyester resin) and colour (Figure 22.2) was prepared, in that catalyst was also added and applied on the wax painted mould structure. On that layer of glass-wool mat (Chopped strand mat; Figure 22.3) and jali (woven roving; Figure 22.4) one over other were placed. For construction of mould according to size of boat 14-18 glass-wool mat and over that 14-18 woven roving jali are attached with the help of resin.

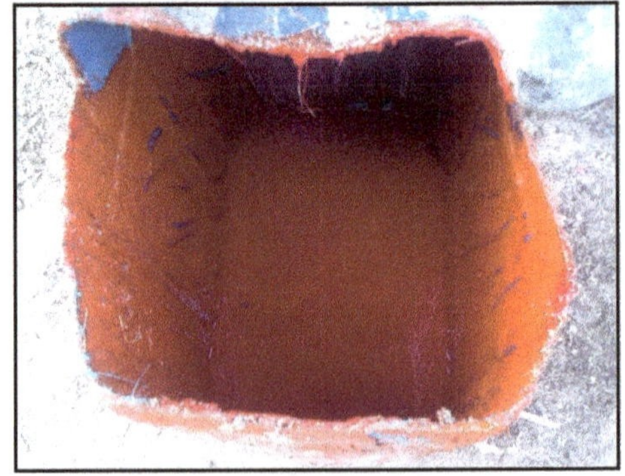

Figure 22.2: Gel-Coat Mixture with Colour.

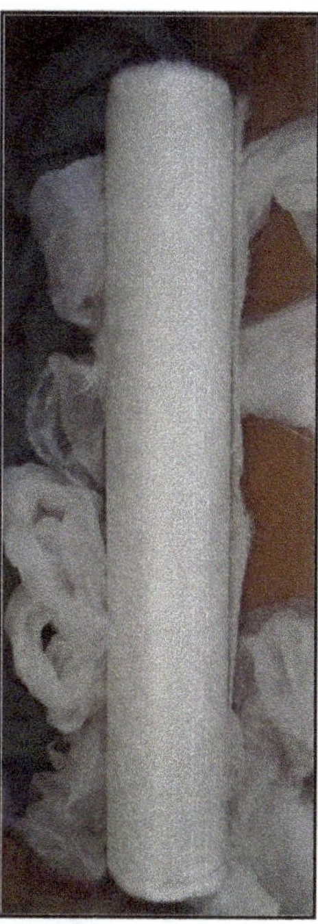

Figure 22.3 Glass-wool Mat (Chopped strand mat).

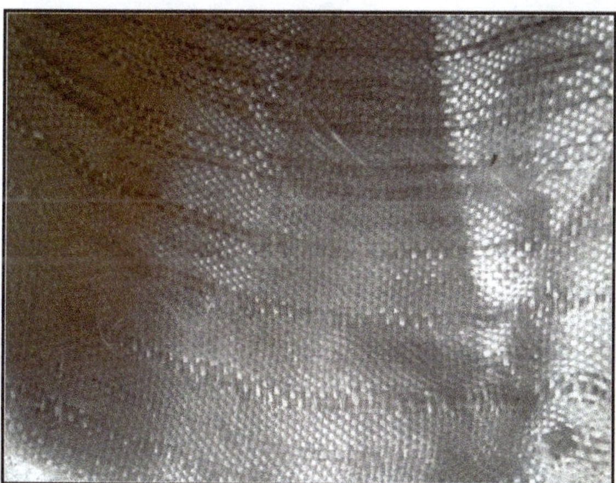

Figure 22.4: Glass-wool Jali (Woven roving).

3. Preparation of "Lapi"

Then mixture of catalyst (*Methyl Ethyl Ketone Peroxide*), *Cobalt naphthanate*, powder (Resin Putty) and colour, was prepared which is called as "*Lapi*" (Figure 22.8). This *Lapi* was then applied on the glass-wool mat and woven roving jali layer. According to required colour, colouration of *Lapi* was prepared. Then allowed that mould structure to dry up to 20 days.

Figure 22.5: Catalyst Methyl Ethyl Ketone Peroxide and Colour.

Figure 22.6: *Cobalt naphthanate* and Resin Putty.

Figure 22.7: Blue Coloured "*Lapi*".

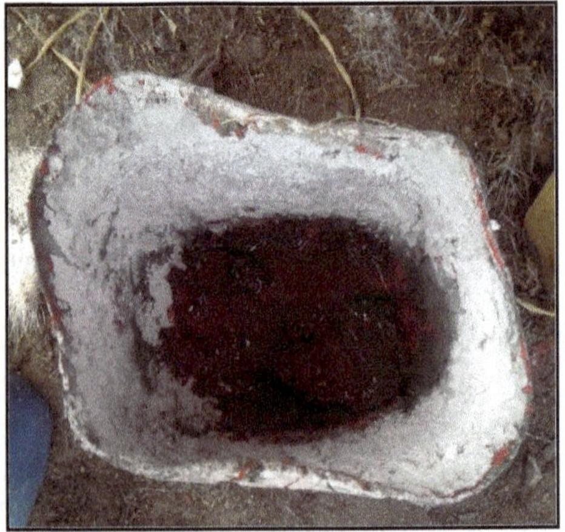

Figure 22.8: Green Coloured "*Lapi*".

4. Fitting of Mould with Keel

After completion of above procedure the keel was prepared of required size. Then the dried structure of mould was removed from the mould skull and attached with keel, which form a shape of trawler (Figure 22.9).

Deck Portion (Stern view)

Deck Portion (Stem view)

Figure 22.9: Boat Construction.

5. Wooden Constructin of Boat (Ribs Fitting, Planking, Deck Fitting, *etc.*)

Then according to the shape of boat wooden construction was done, inside and outside surface of the boat. After keel fitting the ribs were fitted from inner side of the trawler and then planking was done. After deck planking and deck fitting; cold storage, fish storage, machine room and cabin room were constructed (Figure 22.9 and 22.10). For both the storage rooms, fibre or PUF coating was done to avoid temperature exchange. Cabin was made by using wooden material and fitted towards the stem portion.

Figure 22.10: Machine Room.

6. Propulsion

The engine power of the boat ranges between 105-205 hp. (Figure 22.11) and according to engine power propeller size ranged from 38-46 inch. The length of FRP trawler ranged between 42-44 feet and width ranged between 19-20 feet (Figure 22.12).

7. Wooden Material and Expenditure

For boat construction wooden material which used mostly was, *viz.* Sag wood, Sal wood, Babul wood, Saru wood, Nimb wood, *etc.* Up to 2000 kg. wood was required for construction of 42-44 feet boat. For construction up to mould 5-6 Lakh Rs. expenditure was there and for construction of total FRP Trawler up to 25-30 Lakh Rs. expenditure was there. For complete construction of boat 4 to 5 month period was required and also construction was done in period as per owners demand.

Figure 22.11: Boat Engine.

Conclusion

In the coming years, wooden boats will be replaced by FRP fishing boats because these boats are rot free, corrosion free and also stronger. When FRP boat left unused and unattended over the years, it will remain unchanged except for some dulling of the gel-coat shine which may be restored by polishing. Twenty years is often quoted as the lifespan of an FRP vessel. Overall comparison with wooden boats, these boats are more beneficial and economic. Therefore FRP Trawlers will become good option for fisherman.

Figure 22.12: FRP Trawler.

REFERENCES

Coackley, N.1991.Fishing Boat Construction: 2 Building a fibreglass fishing boat. FAO Fisheries Technical Paper. No 321. Rome: FAO.

FAO.2010.Training manual on the construction of FRP beach landing boats, Electronic Publishing Policy and Support Branch, Communication Division, FAO, Viale delle Terme di Caracalla, Rome- 00153, Italy. pp. 13-26.

http: //www.answers.com/topic/frp-1#ixzz2JLE8xfuz.

Laszlo, P. D. D.1960.Glass Reinforced Plastic Hulls. In: Fishing Boats of the World: 2 (Traung, J. O., Eds). London: Fishing News (Books) Ltd.

Shamsuddin, M. Z. 2003. A conceptual design of a fibre reinforced plastic fishing boat for traditional fisheries in Malaysia. UNU - Fisheries Training Programme. pp. 1-53.

Chapter 23

Effect of Quinolphos and Deltamethrine on Respiratory Activity Levels in Freshwater Fish *Channa orientalis*

☆ *R.D. Patil*

ABSTRACT

Quinolphos and Deltamethrine on respiratory activity of freshwater fish Channa orientalis was studied after acute (24 to 96 hrs.) and chronic (7 to 21 days) treatment. After acute (5.2115 ppm concentration of Quinolphos and 4.1043 concentration of Deltamethrine) and chronic (1.042 3ppm of Quinolphos and 0.8208 ppm conc. of Deltamethrine) treatment the rate oxygen consumption was found to be decreased.

Keywords: *Quinolphos, Deltamethrine, Respiratory activity, Channa orientalis.*

INTRODUCTION

Fishes are known to be the richest source of protein. Fish are exposed to aquatic toxicants through their gills, which are multifunctional and complex organs with which they make intimate contact with the ambient water, and as in case of sea water fish via drinking (Wendelaar Bonga,1997). Various kinds of biological effects of different environmental toxicants have been studied in a variety of fish (Jyothi and Narayan,1999; Moore and Waring, 2001; Fulton and Key, 2001). Fish are widely used to evaluate the health of aquatic ecosystems because pollutants build up in the food chain and are responsible for adverse effects and death in the aquatic systems

(Farkas *et al.*, 2002; Yousuf and El-Shahawi, 1999). Oxygen consumption rate of freshwater fishes differs from species to species. Quinolphos and Deltamethrin are toxic pesticides to aquatic organisms which might be hampering fish health through impairment of metabolism sometimes leading to death. Many species of animals are able to take oxygen from water as well as from air. Investigations on the respiration of air breathing fishes have been made by many investigators such as Albaster *et al.* (1957) and Banerjee and Sen (1971). Apparently no attempt has been made to determine the exact relationship between oxygen consumption and activity of fishes hence the present investigation was undertaken to study the effect of Quinolphos and deltamethrin on respiratory activity levels in freshwater fish *Channa orientalis.*

Materials and Methods

The freshwater *Channa orientalis* were collected from Tapi river at Vanjari, Taluka Uchchal, District Tapi (G.S.), India. The specimens were collected from their natural habitats, brought to the laboratory and acclimatized to laboratory conditions for a week. Healthy and same sized fishes (length 9-10 cm and weight 37.50±0.70 g) were used for the experiment. The fishes were fed with standard powdered feed and starved for 24 h prior to the experimentation. The fishes were divided into two groups, one group was kept as control and second group was exposed to 5.2115 ppm concentration of Quinolphos and 4.1043 conc. of Deltamethrin for 24, 48, 72, and 96 h as acute treatment. For chronic treatment fishes were exposed to 1.0423 ppm of Quinolphos and 0.8208 ppm conc. of Deltamethrine for 7, 14, and 21 days. The amount of oxygen was determined by the standard Winkler's method as given by Welsh and Smith (1960). The total oxygen consumption was calculated by considering wet weight of fish and expressed as ml of O_2 consumed/g of soft body wt/hr/lt. The "t" test was carried out and percent change in oxygen consumption was noted.

Results and Discussion

In the present investigation in *Channa orientalis* (Schneider), it has been found that Quinolphos and deltamethrin intoxication brought significant decrease in oxygen consumption as compared to control, a view consistent with the finding of Gopalakrishna Reddy and Gomathy (1977) in thiodon exposed *Mystus vittatus,* Vasanty and Ramaswamy (1987) in thiodon exposed *Sarotherodon mossambicus* and Kumar *et al.* (1998) in metacid exposed *Heteropneustes fossilis.* The rate of oxygen consumption by *Channa orientalis* is summarized in the Table 23.1 and 23.2. Respiration controls the metabolic activity. Oxygen is essential to provide energy for life processes and its requirement vary according to the activity of the fish. As weight of fish increases initially it showed sharp decline in rate of respiration and in due course of increase in weight of fish there was extremely sharp increase in the rate of O_2 consumption up to certain level and beyond this the rate of oxygen consumption decreases even though there is increase in the body weight (Pampatwar and Ambore,-2004). Pesticides damage the gill surfaces and reduce the oxygen uptake capacity of the respiratory organs (Roberts,1978). The exact reason of decrease

Table 23.1: Rate of O_2 Consumption of *Channa orientalis* after Acute Exposure to Quinolphos and Deltamethrine

Sl.No.	*Treatment*	*Average O_2 Consumption + S.D. ml/g/hr/lt*			
		24h	*48h*	*72h*	*96h*
1.	Control	35.832 ± 3.421	35.801 ± 2.791	35.798 ± 2.604	35.700 ± 2.537
2.	Quinolphos	31.915^{NS}± 2.058 –10.93	30.241*± 1.933 –15.53	28.793*± 1.569 –19.56	25.114**± 1.180 –29.65
3.	Deltamethrin	29.041*± 2.146 –18.95	27.348*± 1.6099 –23.61	25.652*± 1.375 –28.34	20.423**± 1.052 –42.79

Each value is mean of three observations ± S.D.

(+) or (-) indicate percent variation over control.

Values are significant at *= $P < 0.05$, ** = $P < 0.01$, *** = $P < 0.001$. NS = Not Significant.

Table 23.2: Rate of O_2 Consumption of *Channa orientalis* after Chronic Exposure to Quinolphos and Deltamethrine

Sl.No.	*Treatment*	*Average O_2 Consumption + S.D. Ml/gm/hr/lt*		
		7D	*14D*	*21D*
1.	Control	35.617± 2.447	35.491± 1.708	35.012± 1.608
2.	Quinolphos	23.988**± 1.145 –32.65	21.217***± 1.065 –40.21	20.461***± 0.977 –41.56
3.	Deltamethrin	18.171***± 1.004 –48.98	17.233***± 0.922 –51.44	15.368***± 0.811 –56.10

Each value is mean of three observations ± S.D.

(+) or (-) indicate percent variation over control.

Values are significant at *= $P < 0.05$, ** = $P < 0.01$, *** = $P < 0.001$. NS= Not Significant.

in oxygen consumption in *Channa orientalis* after pesticidal treatment in the present study could not be understood and it needs further investigation. However it may be considered as an adaptive mechanism to avoid the stress of pesticide in the aquatic media, a view consistent with the finding of Kumari and Pandey (2009).

REFERENCES

Albaster, J. S., Herbert, D.W.M. and Hemens, J. 1957. *Ann. Appl. Biol.* 45 (1): 177-188.

Banerjee, S.C. and Sen, P. R. 1971. *Sci. Calt.* 35(9): 1970.

Farkas,A., Salanki, J.; Specziar, A.2002. Relation between growth and the heavy metal concentration in organs of bream *Abramis brama* L. populating lake Balaton. *Arch. Environ. Contam. Toxicol.*,43(2): 236-243.

Fulton, M. H. and Key, P. B.2001. Acetylcholinesterage inhibition in estuarine fish and invertebrates as an indicator of organophosphorus insecticide exposure and effects. *Environ. Toxicol. Chem.* 20, 37-45.

Gopalakrishna Reddy,T.and Gomathy, S. 1977. Toxicity and respiratory effect of pesticides, thiodon on catfish, *Mystus vittatus, Indian J. Environ Health,* 19 (4): 360-363.

Jyothi B., Narayan, G,1999. Certain pesticide induced carbohydrate metabolic disorders in the serum of freshwater fish *Clarias batrachus* (Linn). *Food Chem. Toxicol.* 37: 417-421.

Kumar Arun; B.N.Pandey; Arshi Rana and Ranjan Kumar 1998. Toxic effect of metacid on bimodal oxygen uptake in an air breathing fish *Heteropneustes fossilis* (Bloch). *Columban, J. Life Sci.* 6(2): 339-341.

Kumari, Poonam Rani and Pandey, A.K.2009. Effect of Taficide (Herbicide) on bimodal oxygen uptake mechanism in an air breathing siluroid fish, *Heteropneustes fossilis* (Bloch). *Nat. J. Life Sci.* 6 (3): 359-360.

Moore, A. and Waring, C.P.2001. The effect of a synthetic pyrethroid pesticide on some aspects of reproduction in Atlantic salmon (Salmo salar L.). *Aquat. Toxicol.*52, 1-12.

Pampatwar, D.V. and Ambore, N.E. 2004. The respiratory activity levels in a sluggish and active fish A comparative study from Nanded District, Maharashtra. *J.Aqua. Biol.*19(2): 93-100.

Roberts, R.I. 1978. Pathophysiology and systematic pathology of teleost In: Fish Pathology (Ed. R.IO.), pp 55-59.

Vasanthi, J.S.D. and Dube,S.C.1973.Oxygen uptake capacity of gills and skin in relation to body weight of the air breathing fish, *Anabas testudineus* (Bloch). *Acta.Physiol. Acad. Sci. Hung.* 44: 113-123.

Vasanty, M. and Ramaswamy, M. 1987. A shift in metabolic pathway of *Sarotherodon mossambicus* (Peters) exposed to thiodon (endosulfan). *Proc. Ind. Acad. Sci. (Anim. Sci.)* 96 (1), 55-61.

Welsh, J.H. and Smith R.I.1960.Laboratory exercise in Invertebrate Physiology. Burgess minneapolles.

Wendelaar Bonga,S.E.,1997. The stress response in fish. *Physiol. Rev.* 77: 591-625.

Yousuf, M. H. A.; EI- Shahawi.1999. Trace metals in *Lethrinus lentjan* fish from Arabian Gulf. Metal accumulation in kidney and Heart Tissues. *Bull. Environ. Contam. Toxicol.*, 62 (3): 293-300.

Index

E

F

G

H

I

J

L

M

N

Priacanths hamrur (p. 33)

Decapterus russelli (p. 134)

Alectis indicus (p. 134)

Alepes djedaba (p. 134)

Megalaspis corydyla (p. 134)

Caranx rotteleri (p. 36)

Terapon jarbua (p. 36)

Johnius vogleri (p. 37)

Johnius aneus (p. 37)

Lactarius lactarius (p. 39)

Leiognathus dussueri (p. 39)

Pampus argenteus (p. 38)

Secutor ruconius (p. 40)

Secutor insidiator (p. 40)

Rastrelliger kanagurta (p. 41)

Scomberomous lineolatus (p. 41)

Mugil cephalus (p. 42)

***Perpheris moluca* (p. 43)**

***Sardinella longiceps* (p. 43)**

***Saridinolla fimbriata* (p. 44)**

***Escualosa thoracata* (p. 44)**

***Dussumieria acuta* (p. 44)**

***Opisthopterus* sps. (p. 45)**

***Thryssa malabarica* (p. 45)**

***Thryssa setirostris* (p. 46)**

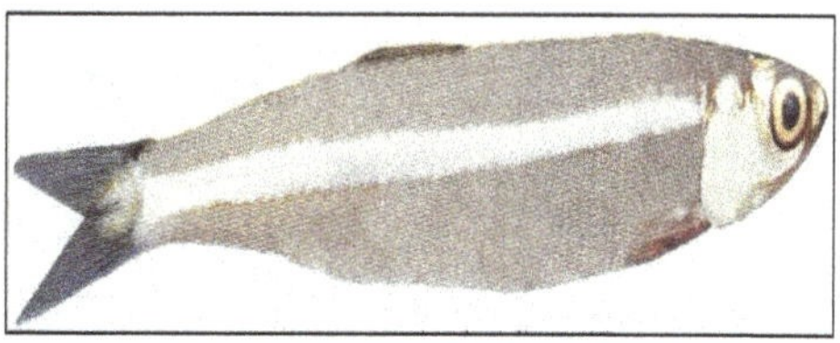

***Anchoviella indicus* (p. 46)**

Stolephorus devisi (p. 46)

Arius sps. *(p. 47)*

Cynoglossus macrostomus (p. 48)

Cynoglossus puncticeps (p. 49)

Heniochus acuminatus, Wimpal Fish or Banner Fish (p. 102)

Chaetodon collare, Butterfly Fish (p. 103)

Pomacanthus annularis, Bluering Angle Fish (p. 104)

Odonus niger, Blue Trigger Fish (p. 105)

Acanthurus mata, Surgeon Fish (p. 105)

Gymnothorax favagineus, Laced Moray Eel (p. 106)

***Ambassis commersoni*, Glass Fish (p. 106)**

***Neopomacentrus filamentosus*, Moon Tail Damselfish (p. 107)**

***Etroplus suratensis*, Pearl Spot (p. 107)**

***Scatophagus argus*, Spotted Scat (p. 108)**

Platax teira, Teira Batfish (p. 109)

Monodactylus argenteus, Silver Moony (p. 110)

Figure 7.1: Hatchery Set-up for Discus Breeding. (p. 116)

Diodon hystrix, Spotfin Porcupine Fish (p. 110)

Figure 7.2: Red Passion Discus Pair with Eggs on Breeding Cone. (p. 116)

Figure 7.3: Discus Pair with Babies on Breeding Cone. (p. 117)

Figure 7.4: Parental Care. (p. 117)

Figure 10.1: *Tilapia mossambica.* (p. 137)

Figure 7.5: Leopard Discus. (p. 118)

Figure 7.6: Pigeon Discus. (p. 118)

(*All Figure Courtesy*: Mr. Alhad Phadnis, Pune)

Figure 10.2: *Cyprinus carpio.* (p. 141)

Figure 10.3: *Clarias gariepinus.* (p. 143)

Figure 10.4: *Ctenopharyngodon idella.* (p. 145)

Figure 10.5: *Carassius auratus. (p. 146)*

www.ingramcontent.com/pod-product-compliance
Ingram Content Group UK Ltd.
Pitfield, Milton Keynes, MK11 3LW, UK
UKHW021010290726
14059UKWH00001BA/52

9 789386 071484